Engineers at Work: A Casebook

ENGINEERS

AT WORK:

HOUGHTON MIFFLIN COMPANY   BOSTON

Atlanta   Dallas   Geneva, Illinois   Hopewell, New Jersey   Palo Alto   London

A Casebook

KARL H. VESPER

University of Washington

To my ladies—Joan, Karen, Linda, Holly, and Nancy

Stanford University has kindly granted permission to re-
produce the following cases:

Dick Rigney (ECL 8), Jack Wireman (ECL 14), Warren
Deutsch (ECL 25), J. L. Adams (ECL 36), Irv Howard
(ECL 58), Art Whiting (ECL 60), Ed Fish (ECL 73),
Hendrik Van Ark (ECL 77), Keith McFarland (ECL
112), and Terry Abbey (ECL 122).

Printed in the U.S.A.

Library of Congress Catalog Card Number: 74–15590

ISBN: 0–395–18407–X

Contents

Preface

Compared to other fields, such as law and business administration, use of cases is still relatively new and limited in engineering. It is my hope that this book will help make the case approach more widely available and its advantages better known. Cases are samples — specimens from industrial experience — and in contrast to more conventional textbook materials, they can be exploited in a greater variety of ways. Although suggestions will be offered here for ways to employ them based on prior teaching experience, it may happen that some users will break fresh educational ground with previously untried approaches. I hope the results of such pioneering by users will be disseminated so that they will be available to myself and others.

The work of many individuals has helped make the present form of this casebook possible. For introducing me to the craft of casewriting I owe most to Andrew Towl and the late Arnold Hosmer, both of the Harvard Business School. Among early encouragers of transfer of the case method from business to engineering, I would like to recognize especially James L. Adams, the late John Arnold, William Bollay, Peter Z. Bulkeley, Henry O. Fuchs, Newman Hall, William Kays, Stephen Kline, Richard K. Peffley, Joseph Pettit, William Schimandle, Robert Steidel, and Frederick Terman.

Preparation of individual cases in this book also has benefited from direct contributions by many people. Each case was made possible by the cooperation and materials from one or more industrial firms or other organizations. The DATA International case was written by Professor James Hill of the California State University at Long Beach; the Colonel Jack Bristor case by Professor Jack Bristor of the University for Florida; the Bob Knowlton case by Professor Alex Bavelas of the University of Victoria; and the Laboratory Equipment case is reproduced through permission from the Harvard Business School. The Flight Safety Foundation kindly allowed use of the design notes appearing in that case. The remaining cases are reproduced by permission from Stanford University, which in turn was supported in their preparation by grants from the National Science Foundation.

Professor Ralph Smith of Stanford University helped in preparation of the Ed Fish case as did casewriter Susan Hays. Other casewriters who contributed include Eugene Echterling, Charles Fernald, David Horine, and Mitchell Blanton.

Thanks are also due to Houghton Mifflin for their encouragement and suggestions on the book, and to reviewers whose comments were most helpful, particularly Professor Godfrey Savage of the University of New Hampshire.

With all this assistance it should probably have been possible to bring this book closer to perfection. For the extent to which that was not done, the responsibility must remain with me.

Karl H. Vesper

Engineers at Work: A Casebook

One   USE OF CASES

To bridge the gap between abstract concepts of engineering science and the practitioner's complex world, with its nonideal problems, unorthodox procedures, and human fabric, is a difficult but important challenge for engineering schools. In selecting courses of study — and beyond that, career patterns — students like to know what the options are and what they entail. Industry at the same time has indicated its preference that students graduate with some appreciation of what actual practice will involve and how it utilizes academic theory.

What is the work of a practicing engineer like, and how can school training and self-development relate to it? How does better versus poorer engineering practice compare or contrast with the student's familiar experience of right versus wrong answers on math and physics problems? To what extent does the way theory and practice interact depend on circumstances of a particular engineering situation, and what is the nature of that dependence? Can poor theory produce a good product, or good procedures a poor product, and if so, how? What can or should an engineer do about it? How should the capabilities of a trained engineer differ from those of a person who is simply "good at making things that work"?

"It depends" will often be a suitable response to such questions, and in order to explore the dependence it will be helpful to have available some "for instances." These could take many forms — stories, movies, books or direct experiences. Written cases are the form they take in this book. Each describes in words and pictures an actual problem situation from industry so the student can examine circumstances engineers face. The cases are designed so that the student can struggle with problems engineers confront, both specifically and generally.

Cases represent only one out of a number of approaches for bridging the gap between theory and practice. Each approach has its strengths and shortcomings, and it makes sense to consider how several may be combined to complement one another, rather than to debate which one is best.

Books of the conventional theory-plus-problem-sets type generally concentrate on idealized problems to clarify theory. This is a difficult way to learn, but it is vital. It also requires idealizations which compromise reality. Frictionless bearings, point loads, and isentropic processes never apply as perfectly in practice as they do in the answer books. Practical judgment must also be added.

Books relating the history of engineering achievements, such as solving the Lockheed Electra crashes* and the development of the V-2 rocket,† often effectively present drama and human events of engineering work, but usually — presumably to suit mass markets — omit technical details. Technical reports, on the other hand, do the opposite, often omitting personalities entirely. They also usually dwell on solutions rather than problems or the process by which they were worked out.

Lecture anecdotes from the industrial experiences of faculty or visiting speakers, to the extent they are available and fit the needs of a particular course, have the advantage of being first hand. They can also include both technical and human details. A disadvantage, depending on how the stories are presented, can be that students may be so intimidated by the superior knowledge and authority of the speaker that they will listen passively, as to a television, rather than entering the fray to struggle themselves with possible solutions to the engineering problems. By only listening rather than participating, they will derive less from the experiences presented.

Field trips to local industrial organizations allow students to touch, see, hear, and smell the processes of manufacturing the end products of engineering. They also present the possibility of meeting some engineers and viewing their activities (excluding perhaps those important activities of engineering that occur inside the head). What cannot generally be obtained on field trips is sustained exposure to a given program as it progresses. Usually such visits

* Robert J. Serling, "The Electra Story" (New York: Doubleday, 1963).
† Walter Dornberger, "V-2" (New York: Ballantine, 1954).

allow only brief watching, not mental involvement in the problem-solving processes taking place.

Summer and co-op programs allow much deeper immersion in the process, although they often are restricted as to level in organizational hierarchy and stage in the design process. The number of hours of direct exposure to engineering in action can be tremendously greater than in any academic course, and can increase its impact. At the same time, however, much of the activity may be rather routine and not highly educational, since it is not, for the most part, designed to be. Finally, such alternatives are often limited to a minority of students, and even for those to only one or two different jobs in the course of school.

Projects in school have the advantage of faculty supervision. They can allow students great creative freedom to design something with their own imprint on it, a refreshing contrast to the discipline of problem sets. If time allows and the task is modest enough, students may even be able to build and test a jury-rig, which adds to the adventure of obtaining an outcome plus the instructional value of nature's verdict. Unfortunately, however, it will not be possible in a school term to do many such projects, perhaps not more than one. Also excluded will be the kinds of feedback that lie downstream from prototype trial, such as debugging, production design, labor, other fabrication difficulties, and finally customer trial.

Written cases such as those appearing in this book also have strengths and weaknesses, which let them complement other techniques such as those mentioned above. The aim of a case is to present a real engineering-problem episode in compact written form so a student can project himself into the story and perform some of the work needed. The problems may not always be explicitly defined and may be part of a background in the case, which also introduces personalities, conveys views of the real engineers, and presents samples of their notes illustrating how they work. Part of the student's job may first be to ascertain what problems exist, and then to define them as a prelude to solution.

Cases seek to combine some of the technical detail, which might appear in normally impersonal reports, with personal viewpoints, which appear in nontechnical histories. Both ingredients are present in real problem solving. In this way cases can have some of the advantage of first-hand anecdotes. They inevitably lack the presence of the living personality telling the story, but they have the advantage of being available at the user's convenience, of being static to allow sustained examination, and of being structured to allow participation in problem solving by the reader. They can be more compact in terms of time than most projects, because they can focus on key events and decisions, skipping over much tedium to present high points of engineering ex-

perience on a collapsed time scale. They must to some extent sacrifice the open-endedness that allows free creative play in projects, but they can offer glimpses of engineering experience far downstream from the point where school projects must usually end, such as the problems that may arise after customers begin using a new design. Through a series of such cases a student in one term may be able to work at different stages — from preliminary design to field test — in several major projects of different industrial organizations.

By combining cases with other teaching techniques, variety can be increased while different perspectives on engineering are also made possible. One combination, for example, can include a series of cases in a design course along with a term design project, using the cases to illustrate how projects in industry may vary from the nature of the ones used in the course, while the course project itself becomes an indepth effort in one particular direction. In combination with other texts and problem sets, cases can offer a break in the pattern as well as illustrate the interfaces between theory and practice. This case book is thus intended to complement other books and teaching methods rather than to completely replace them.

Level of Difficulty

The academic level for which cases in this book were originally written ranged widely. The Bob Knowlton case was written for graduate students, while the Jack Wireman case was written for machine design seniors. The Terry Abbey case was written for freshmen. But all these cases can be used at all college levels.

Most of the cases pose unsolved technical problems that can be attacked with methods of high-school and freshman physics. All portray some processes of engineering in action, and no technical tools are required to explore these processes in general ways. Looking at and learning from events and processes that are already available tools for understanding are not uncommon human experiences, either in or out of school. The specimens of engineering in this book should be profitably examinable by any student with instructor guidance. Those with greater technical training should be able to perceive and appreciate the cases in greater depth, just as artists tend to notice more in paintings and geologists more in scenery than do others who lack their special training.

It is highly important to recognize, however, that not all students should be expected to be able to deal with every technical problem in these cases. That fact should be part of the lesson of this book. Some of the problems should be assigned for working, and others simply for looking at, depending on prior training. The first, the Bob Knowlton case, has no technical problems. It deals only with human relations, which can be dealt with by almost anyone, although people with human-relations training may see more subtleties. At the other extreme, the Keith MacFarland case gives an analysis of electronics design using highly sophisticated mathematics, which only electronics graduate students might be expected to command. The purpose of the latter case is simply to display a specimen of high-powered theory in action.

Nature of the Cases

Beyond the fact that their usability tends not to be restricted to any particular academic level, cases differ from more familiar learning tools of engineering in a number of other respects, and it is worthwhile to note several distinguishing characteristics. These will have implications for both student and instructor regarding possible conclusions to be drawn from cases, what homework should consist of and how it should be done, how class meetings should operate, and how student performance should be evaluated.

Although chosen with care, the cases herein were not selected to "propagandize" students into particular value judgments about engineering. The cases do not concentrate on unusual glamor either in terms of hardware worked on by the engineers or in terms of organizational positions of the engineers. Nor were the cases selected to present either "good" or "bad" examples of engineering performance. Because it is a human endeavor, engineering includes all types of activities, positions, performances, and outcomes, and it is often possible in hindsight to criticize the way things were done. Exigencies of the real world call upon engineers to operate, at times, outside their individual specialties, and also beyond theory and prior practice. Their procedures are not always as precise and straightforwardly methodical as some might wish.

Students witnessing such unglorified human episodes may at first be inclined to be quite critical of some engineers presented in the cases, for ways they have proceeded. But when the students undertake the problems and try to do better, they may more fully appreciate strengths exhibited by the engineers, as well as learn from the trial-and-error processes of both the professionals and themselves.

The cases refrain as fully as possible from expressing opinions except when they are quotations or paraphrasings of comments by characters in the cases. To convey facts of the case situations, use is made of textual information, excerpts from engineers' notes, catalogs, drawings, and photographs. The aim of these is to present the reader with as accurate a portrayal as possible of the actual circumstances faced by participants in the actual engineering events, so he can identify with the characters and attempt to solve some of the problems himself.

People in the cases will at times offer opinions about their particular circumstances, or engineering in general, which they held at the time of the case. These may later be viewed as right or wrong depending on how the reader reasons. But it is important to note that opinions expressed by personalities in the cases should not be presumed to be correct to any greater extent than they would have been accepted if the reader had personally heard them expressed by the case characters at the time they were spoken or written. The ability to sift and filter fact from opinion is something students should practice in dealing with the cases presented here.

Clearly, any attempt to portray reality after the fact — limited as we are to printed pages — will suffer from imperfections. Readers cannot smell the steel-mill effluents Terry Abbey must dispose of, hear the scream of the failing motor in the Jack Wireman case, watch the ominously rising flood waters of the Detroit River as Jack Bristor worries what action to take, or feel the pushbuttons on Warren Deutsch's satellite control panel. Some aspects of the real thing will always be missing in a written description.

One important missing ingredient is the ability to find more information about the real situation. Students who decide they want to know more facts about the circumstances of a case may be able to search in the library or telephone a supplier for certain portions, perhaps, but they cannot run another test on Jack Wireman's bearing or talk to the union people who limit Art Whiting's freedom to putter in the shop. A ground rule for working the cases should consequently be that students must do the best job of analysis, creative thinking, and decision making they can with the information they have available. This is not an unrealistic requirement, because information is not infinite in the real situation either, and even the working engineer may often wish he had more data before making a decision. Pressures of time and money may still limit him from obtaining more information. It has been said that part of the art of engineering is making good decisions in the absence of complete information. In school this art can be practiced with minimal penalty for error.

Lack of demonstrably "right" answers can be another frustrating aspect of cases, just as it is of design projects. Unlike a typical application-of-science or closed-form mathematical problem, where it may be possible to show by irrefutable logic following from established theoretical postulates that an answer is absolutely "correct," it is generally not possible to do so with case problems. Engineering solutions in real life are continually being improved, showing that an answer which was accepted as "ultimate" before was not really the best that could be done. When working on problems, engineers sometimes find solutions they really do not understand. Under time pressure to correct a machine failure, for example, many reme- dies may be tried at once. If a cure happens to re- sult it may not be possible to say which remedy did the job. The engineer may not, moreover, be able to delve further into the causality, but rather may simply have to get on to the next job with no time for more than a brief thought of relief that some- how the solution worked.

It is also possible to have successes that prove to be failures and failures that prove to be successes. One toy company put a great deal of sophisticated technical analysis into the design of a better walking doll. To avoid the jittering mechanical gait of former products, a careful study was made of the interaction between the rocking of the doll's body and movement of the legs. A doll resulted that walked with impres- sive naturalness. Unfortunately, it would only move forward if given a shove, otherwise it would only stand in one spot and gyrate. Further study and experiment revealed that the doll would start forward after a few cycles if the soles of its shoes were fitted with pliant material to allow more buildup of rocking energy. Further tests revealed another problem. The doll could not walk on carpets. Still more study and experimentation was done, and from this it was found necessary to articulate the doll's ankles. Now it would start and walk on or off carpets, even roller- skate with an amazingly lifelike gait. Upon its in- troduction to the market the toy was widely ap- plauded and won an industrial prize for innovative design. At this point, it was judged a success.

But the market gave a different verdict. At a market price of around $30, which was necessary to cover manufacturing costs, parents simply did not buy the product, wonderful as it was. So the ap- parent success turned into apparent failure.

A similar verdict was placed by observers on an- other engineering innovation, an oceanographic data buoy developed for the Navy. Many millions of dol- lars were spent on developing this buoy, which was to be anchored in the open ocean and telemeter real-time data on variables such as water current, temperature, and waves, as well as on meteorological properties of the air. It was expected that the data would be useful both for keeping track of Russian submarine move- ments and weather forecasting.

It turned out, however, that submarines or other vessels could easily counteract such surveillance sim- ply by sinking the buoy, stealing it, or cutting it adrift. Weathermen, on the other hand, did not show much interest in the meteorological data, particularly as satellites were already beginning to give them a superabundance of new data. At the same time, each buoy was so expensive and complex that it be- came widely known as the "monster buoy." It also became known as a failure.

Then the idea occurred to someone that although the buoy was expensive and complex, it was con- siderably less so than lightships used along coasts to warn passing vessels of rocks, and it began to appear that the buoy could perform the same neces- sary functions as lightships. So the verdict changed and the monster buoy became known as a successful engineering development after all.

Which answers were "right" and which were "wrong" in these two projects and many others is thus a function of unanticipated as well as expected external events. It can vary with time, and clearly it can be expected that there will be different judg- ments by different people. A success in the eyes of some may be a failure to others. This possibility is particularly evident these days as the role of tech- nology and its social and ecological interactions is debated.

In the absence of absolute answers regarding out- comes, evaluation of student performance must be based much more upon the ways they reason in reaching conclusions. Success must be judged more upon the journey than on its destination. There is, of course, no *the* approach, just as there is no *the* answer, but it can be strongly argued that some ways of working are more likely to be successful than others. Reasoning from available facts usually works better than reasoning from supposition. Tak- ing care in formulating problems usually produces better results than making hasty assumptions about what the problems are and "shooting from the hip." Pausing to generate several alternative solutions for an important problem before deciding which to pur- sue is often more advisable than falling in love with the first idea for a solution that comes along.

When parts of the problem offer opportunity to apply mathematical techniques or engineering sci- ence, it is obviously preferable to perform quantita- tive analysis correctly and put the decimal point in the proper place, as opposed to making wild guesses or simply ignoring quantitative data in favor of ver- balizing. Sometimes it makes sense to round off quantities and make major simplifying assumptions, and other times it is better to work in great detail. In drawing conclusions about what depth of analy- sis to apply, as well as in making decisions based upon that analysis, it is often possible to distinguish between connected and specious reasoning. In short, right or wrong design answers notwithstanding, it is possible to distinguish better and poorer engineering processes for arriving at answers.

Regrettably, no guarantee of a happy outcome goes with the best procedures at all times, but a higher probability of success does. Just as in Las Vegas, it is possible to bet with the odds and lose or to bet against them and win. But in the long run it is the better procedures, like betting with the odds, that tend to produce more favorable payoffs. Hence it makes sense in school to place higher emphasis on how the student approaches and proceeds on a case than on the solution with which he or she ultimately concludes.

The outcome produced by the practicing engineer in a case may or may not be what the student will produce, depending on whether he and the student follow identical courses of action, which is highly unlikely. For the student it would be comforting to know how the real engineer's answer turned out. But at the same time a display of the case engineer's answer may tend to shift the emphasis away from how the student operated, or to invoke an invalid comparison between the student's and the case engineer's work. Hence it was with mixed feelings that two types of cases were chosen for inclusion in this book, one that includes outcomes from the real project and a second type that does not.

Two Types of Cases

The Jack Wireman case appears throughout the book in a series of parts. Each part ends with an unsolved problem, as it really occurred at the time. The succeeding part then tells what Mr. Wireman did and what events ensued. It is possible, of course, to start from the final outcome and read toward the front with "20/20 hindsight," and this may be tempting because it is easier than struggling with the problems and prescribing action *before* the outcome as the real engineer had to do. The reader must decide for himself what approach will benefit him most.

Other cases that, like the Jack Wireman case, contain both problem and outcome parts, appear in Chapter 7 in the serial-history cases. It should not be assumed that these outcomes are any final word. Beyond them there will always be additional outcomes that can, as observed above, reverse verdicts. The outcomes will not always be satisfactory explanations, for they tell only what was known at the time, and this may not be all the case engineers or readers might like to know. Because the cases are aimed at closeup views of engineering projects, they also tend to tell only about pieces of projects rather than the broader scope of total projects. There can therefore be a mismatch between the degree of success of the piece and the successfulness of the total project. For broader looks at total projects in less detail, the reader is, again referred to descriptions in historical books mentioned earlier.

Aside from the Jack Wireman case and those of Chapter 7, the rest of the cases in the book are single-chapter-problem cases, that is, the complete case is presented as a unit in a single chapter. Each of these ends with one or more unsolved problems. These should concentrate student practice on the process of arriving at answers in order to develop working and thinking habits effective in solving en-gineering problems. With one exception, the dimensioning problem that appears in the Dick Rigney case, these single-chapter cases have no provably "right" answers.

Case Discussion Questions

At the end of each case part are several questions for consideration and discussion about the case. The instructor may wish to suggest different questions or have students formulate their own. The process of deciding what questions to ask may at times be a most important part of the learning process. The questions provided in this work may be particularly useful to those students who are new to cases or who otherwise have difficulty finding a technique for beginning the case-analysis process.

Traditional Engineering Fields

Trying to relegate cases to conventional engineering specialties such as mechanical, electrical, and civil engineering is made difficult by the nature of the engineering work of industry, which often tends to be interdisciplinary. Thus Terry Abbey — nominally a civil engineer — is worrying about mechanical problems in a pumping system, and Jack Wireman — nominally a mechanical engineer — is working on problems of an electric motor. Many of the engineers, moreover, find themselves embroiled in problems outside all technical engineering fields, as when Art Whiting has to worry about union constraints and Jack Bristor's problems lead him close to politics.

It is usually possible, however, to sort out engineers according to the main fields in which they are operating, even though these may at times be different from the fields in which they were trained and the nature of their work often forces them to make excursions from these main areas of occupation. Table 1.1 attempts to sort the cases included in this book according to main fields of interest in each case.

It is hoped that the mix included herein will present an adequate range of examples, while at the same time allow exploration of the common features of practice that run through all disciplines. Those who want to add further specimens of other engineering specialties may be able to find them in the Engineering Case Library at the Stanford University Engineering School, Stanford, California 94305. This library makes available examination copies of cases desired plus an annotated bibliography describing those available without charge. Additional copies for student use may be purchased.

Chapter Topics

Chapters 2 to 7 are used for grouping cases by the following subjects:

2   Problem Formulation
3   Conceptual Design
4   Technical Analysis
5   Decision Making
6   Outcome Review
7   Serial-history Cases

Each of these chapters begins with a brief discussion of the chapter topic. This material is not meant as a replacement for full treatments available in non-case books in much greater detail, but rather as a general summary and reminder. *Exercises* following the text are also supplementary and intended to allow exploration of topics and practice of skills independently of the cases. They can be used to vary the pattern of case analysis and also to explore in more general ways subjects that can be treated with reference to specific circumstances in a given case. Some questions will suggest exploration of possible generalizations based on patterns that may appear in several cases. Some also call for seeking data beyond this book and its cases.

Groupings of cases under specific chapter topics were made on the basis that those particular cases seemed particularly well-suited to exercising concepts of the topical focus. This was not meant to imply either that problems presented in the cases of a given chapter relate only to that chapter, or that progress in solution of the problems necessarily follows the order of chapter topics. Impressions that engineering follows some sort of systematic, scientific method are not supported by experience, although there is a considerable tendency to follow precedent

and build directly from prior, similar solutions. In some fields, as can be seen in the Hendrik Van Ark case, the following of precedent becomes quite routine in the form of handbook procedures.

Cases in Chapter 2, Problem Formulation, present situations in which something has gone wrong, and the main questions concern what should be done to prevent future difficulty of the same kind. Clearly this is not the only sort of situation in which problem formulation is necessary. For example, seeking out needs involves problem formulation, as does creation of new designs, even though nothing may have gone wrong. Cases and exercises of the latter type appear in later chapters.

Chapter 3, Conceptual Design, offers specific opportunities for creative imagination to generate ideas for new solutions. These can be used for short ideation exercises on which to practice some of the semiformal "creativity tricks," such as brainstorming, checklisting, synectics, and morphological analysis. They can, in some instances, also be used to launch design projects on general concepts suggested by others such as appear in the J. L. Adams case. Opportunities for practice of "cerebral popcorn" need not be limited to these case examples but can be applied in many others as well.

Several cases grouped in Chapter 4, Technical Analysis, call for mathematical application of engineering science, ranging from elementary dimensioning of tolerances up through application of principles of statics and $F = Ma$. These problems can be treated in either elementary or more sophisticated ways at the discretion of the instructor and according to the academic experience of his students. The aim of this section is not in any sense to summarize engineering science or to introduce any new analytical techniques, but simply to illustrate how the need for application of such techniques arises in practice, as

Table 1.1

CASES	TRADITIONAL ENGINEERING DISCIPLINES				
	Aeronautical	*Civil*	*Electrical*	*Industrial*	*Mechanical*
Terry Abbey		X			
Jack Wireman					X
J. L. Adams					X
Irv Howard					X
Dick Rigney					X
Ed Fish			X		
Art Whiting	X				
Hendrik Van Ark		X			
Laboratory Equipment				X	
The Reverend C. W. Van Dolsen			X		
Flight Safety Foundation	X				
Col. Jack Bristor		X			
Warren Deutsch				X	
Keith McFarland			X		

opposed to their emanating from theory as it does in expository texts.

Cases in Chapter 5, Decision Making, deal with situations in which the available alternatives are fairly definite. The problem is to choose among them in as reasonable and rational a manner as possible. The decision making called for is, by the nature of most engineering practice, not the sort that lends itself very gracefully to probabilistic formulations. Rather, these cases call for perceptive, objective examination of the evidence, some straightforward arithmetic in places, and hardheaded reasoning. Consideration of costs naturally becomes an important factor.

Material in Chapter 6, Outcome Review, raises the issue of retrospect in engineering experience and what can be learned from it as a prelude to the serial-history cases that follow in the last section. Chapter 6 raises the question of what it is that engineers learn from experience, and how. It also opens the question of how far an engineer should go in seeking to explain relationships that govern performance of the things he designs. In other words, to what extent is his job simply to come up with something that works, as opposed to seeking to understand *why* it works.

The last chapter gives Serial-history Cases. Each can be treated as a series of separate parts to be dealt with individually, or as a single case. Each case part can also be allocated similarly to the single-chapter cases to focus on practice of different problem-solving concepts. As noted earlier, the concepts may apply at different stages in virtually all engineering problems, but one convenient allocation of the serial-case parts to the various chapter topics in problem solving is as follows:

Problem Formulation
    Colonel Jack Bristor (Part 1)
Conceptual Design
    Warren Deutsch (Part 1)
    Keith McFarland (Part 1)
Technical Analysis
    Keith McFarland (Part 2)
    Keith McFarland (Part 3)
Decision Making
    Colonel Jack Bristor (Part 2)
Outcome Review
    Warren Deutsch (Part 2)
    Jack Wireman (Part 5)
    Colonel Jack Bristor (Part 3)
    Keith McFarland (Part 4)

Although the chapter topics can be applied to cases in virtually any sequence, the sequence of parts in a given case must generally be followed in order.

Ground Rules

Because cases introduce a mode of instruction unfamiliar to many, it may be helpful to spend some time in the first class meeting on discussion and establishment of ground rules. These should include what sorts of rules students should observe on homework and class participation, what kinds of output are expected of students, what role will be played by the instructor, how learning is supposed to take place, and how student performance will be evaluated. The nature of the ground rules may be expected to vary among courses and possibly to shift somewhat over time in a given course, just as they may in professional employment after school.

The typical expectation in preparation of homework will be that the student should read the assigned case with care, analyze it, and develop recommendations for one or more characters in the case. He may be asked to prepare answers to questions appearing at the end of the case, though these are not necessarily all the questions he should explore. He may also be asked to review and appraise critically the performance of one or more characters in the case, and suggest how they might better have done their work.

There will usually be technical problems in the case, which the student can take up and struggle with himself. If the assignment is to take minimal time he may only be asked to define these problems and develop a plan for how to attack them. More often, however, he will be expected to undertake that attack and carry it as far as he can with the information available. In some cases he may lack the training needed for arriving at solutions, and under such circumstances he should not be expected to do so. The instructor may point out these limitations, or he may ask that the student define them for himself and indicate how he would obtain the help he might need if, in spite of his incomplete training, he had to develop solutions anyway.

When information in a case seems insufficient for making necessary decisions, each student should determine for himself what should be done about it. He should first reach the best conclusions he can with the information that is available, indicating what assumptions he needed to make. He may also indicate additional information desired to increase confidence in decisions. If he suggests as a part of his action plan that more information be obtained, he should include a description of how, in the actual case situation, he would obtain it, what he would expect it to cost, how he would use it, and why the cost would be justified.

When using serial-history cases it is important to establish a clear ground rule about reading ahead. If the educational emphasis is to be on "Monday-morning quarterbacking" the engineers in the case, then it may be sufficient to read all the parts starting at any point. Yet if emphasis is to be on student practice in problem solving, then a better rule would be to solve the problems of a given part in the serial before reading further. Since all parts are available in the book, this rule can be violated by a reader wishing to do so (except, of course, in the single-chapter cases). Such violation may be detectable if a student is able later to describe only how the case professional dealt with the problem, and is not able to describe the avenues of possible solution he explored before deciding upon one that might or might not be the same as the one chosen by the case professional.

After students have prepared their homework on the day's case and arrived in class, the most frequently used method of dealing with the case is through open discussion. The instructor may call upon one or more students to tell the rest of the class how they analyzed the case and what they concluded, after which other students should begin to volunteer other views, sometimes supporting and sometimes contrasting with those presented first. Examination of conflicting points of view should bring to light aspects of the case that many in the class will not have noticed and from which the class can therefore learn. Role-playing assignments may at times facilitate emergence of such contrasts.

For such mutual investigation of a case and its possible solutions it is vital that as many students as possible be encouraged to participate. Contributions to class discussion should therefore be recognized by the instructor in grading. To encourage participation the instructor should also make it a practice to call upon students who tend not to raise their hands, as well as those who do. Each student should prepare himself before class with an analysis he can present if called upon, and during the class he should continually perform the demanding task of simultaneously listening to what others are saying, comparing and assessing his views against theirs, and reformulating a new position relative to where the discussion is at each moment so he will be ready to respond if called upon to comment.

Such participation can be an intimidating experience for students until they become accustomed to it, particularly for those who tend to be rather quiet or are unaccustomed to speaking before groups. A helpful way for them to prepare can be to discuss the case before class with a smaller group to explore alternative points of view and get the feel of talking about the particular case. Beyond that each student should simply work up his nerve and plunge into the discussion.

There is always a temptation, especially if he has not spent as much time as he feels he perhaps should have in preparing the case, to sit back and let others do the talking. Rationalizations for doing so may include such thoughts as "Debate is not my thing"; "I admire the strong silent type"; "I'll let the others make fools of themselves before I do"; "They probably know more than I do about it"; "My ideas may be off base, so I won't chance it"; "Maybe if I just

listen I'll get a better idea"; and "It's easier to remain detached and silent." The trouble with following such impulses is, first, that the discussion suffers from lack of what the silent student might have contributed, and second, that detachment makes the class experience much less vivid and hence less profitable for the silent student. He will soon forget what most of the others have said, but he would not forget what he would have said and how the class would have reacted to it. His own participation is therefore a most vital part of the learning experience, and he should be encouraged to take some effort and risk to obtain it.

Introductory Case: Comments to the Student

The first case in this book, the Bob Knowlton case, is intended to be an ice breaker that makes discussion interesting and easy. It deals not with technical problems but rather with people problems in a technical environment. Although far removed from most engineering theory courses, such people problems are nevertheless among the most important, according to practitioners. So the Knowlton case is not merely a warm-up. At the same time, however, it is one with which anyone can deal regardless of prior training.

As you read the case there will be a temptation to prejudge Fester, or Knowlton, or others in the case. But such judgments as "he shouldn't have done that" are not sufficient. What *should* he have done, and more importantly, what should he do now, and why? You may also find it worthwhile to explore the question of why each character behaved the way he did. What assumptions do you have to make about each character to explain his actions? Try casting yourself in Fester's role, or Knowlton's. What feelings might propel you to do what they did? Given those feelings, what might others do, or how could the system be modified to reduce or solve problems? Other questions you may wish to consider appear at the end of the case.

One thing you should find after analyzing the Bob Knowlton case and the others in this book is that there is room for differences of opinion. Other people will reach conclusions different from yours. To rise in defense of your views may be a natural reaction, and may also be a process that has profit. But you should also try to determine whether in the opposing views there are some considerations you did not think of, and beyond that whether there is profitable learning in your oversights. This process of extracting new insights from others' views in mutual discussion is one engineering students do not have much opportunity to practice in most courses, and so it may at first seem unfamiliar and somewhat strange. Expanded appreciation and ability to deal with the unfamiliar is a teaching goal engineering shares with the rest of education.

BOB KNOWLTON People Problems

Bob Knowlton* was sitting alone in the conference room of the laboratory. The rest of the group had gone. One of the secretaries had stopped and talked for a while about her husband's imminent induction into the Army, and had finally left. Bob, alone in the laboratory, slid a little farther down in his chair, looking with satisfaction at the results of the first test run of the new photon unit.

He liked to stay after the others had gone. His appointment as project head was still new enough to give him a deep sense of pleasure. His eyes were on the graphs before him but in his mind he could again hear Dr. Jerrold, the director, saying, "There's one thing about this place that you can bank on. The sky is the limit for a man who can produce." Knowlton felt again the tingle of happiness and embarrassment. Well, damn it, he said to himself, he had produced. He wasn't kidding anybody. He had come to the Simmons Laboratories two years ago. During a routine testing of some rejected Clanson components he had stumbled on the idea of the photon correlator, and the rest just happened. Jerrold had been enthusiastic; a separate project had been set up for further research and development of the device, and he had gotten the job of running it. The whole sequence of events still seemed a little miraculous to Knowlton.

He shrugged out of the reverie and was hunched determinedly over the sheets when he heard someone come into the room behind him. He looked up expectantly. Jerrold often stayed late himself, and now and then dropped in for a chat, which always made the day's end especially pleasant for Bob.

It wasn't Jerrold. The man who had come in was a stranger. He was tall, thin, and rather dark. He wore steel-rimmed glasses and had on a very wide leather belt with a large brass buckle.

The stranger smiled and introduced himself. "I'm Simon Fester. Are you Bob Knowlton?" Bob said yes, and they shook hands. "Dr. Jerrold said I might

find you in. We were talking about your work, and I'm very much interested in what you are doing." Bob waved to a chair.

Fester did not seem to belong in any of the standard categories of visitors: customer, visiting fireman, stockholder. Bob pointed to the sheets on the table. "There are the preliminary results of a test we're running. We've got a new gadget by the tail and we're trying to understand it. It's not finished, but I can show you the section we're testing."

He stood up, but Fester was engrossed in the graphs. After a few moments, he looked up with an odd grin. "These look like plots of a Jennings surface. I've been playing around with some autocorrelation functions of surfaces — you know that stuff."

Bob, who had no idea what he was referring to, grinned back and nodded, and immediately felt uncomfortable. "Let me show you the monster," he said, and led the way to the work room.

After Fester left, Knowlton slowly put the graphs away, feeling vaguely annoyed. Then, as if he had made a decision, he quickly locked up and took the long way out so that he would pass Jerrold's office. But the office was locked. Knowlton wondered whether Jerrold and Fester had left together.

The next morning Knowlton dropped into Jerrold's office, mentioned he had talked with Fester, and asked who he was.

"Sit down for a minute," Jerrold said. "I want to talk to you about him. What do you think of him?" Knowlton replied truthfully that he thought Fester was very bright and probably very competent. Jerrold looked pleased.

"We're taking him on," he said. "He's had a very good background in a number of laboratories, and he seems to have ideas about the problems we're tackling here." Knowlton nodded in agreement, instantly wishing that Fester would not be placed with him.

"I don't know yet where he will finally land," Jerrold continued, "but he seems interested in what you are doing. I thought he might spend a little time with you by way of getting started." Knowlton nodded thoughtfully. "If his interest in your work continues, you can add him to your group."

* This case was prepared by Professor Alex Bavelas of the University of Victoria and is used with his permission.

"Well, he seemed to have some good ideas even without knowing exactly what we are doing," Knowlton answered. "I hope he stays; we'd be glad to have him."

Knowlton walked back to the lab with mixed feelings. He told himself that Fester would be good for the group. He was no dunce, he'd produce. Knowlton thought again of Jerrold's promise when he had promoted him — "The man who produces gets ahead in this outfit." The words seemed to carry the overtones of a threat now.

The next day, Fester didn't appear until midafternoon. He explained that he had had a long lunch with Jerrold, discussing his place in the lab. "Yes," said Knowlton, "I talked with Jerry this morning about it, and we both thought you might work with us for awhile."

Fester smiled in the same knowing way that he had smiled when he mentioned the Jennings surfaces. "I'd like to," he said.

Knowlton introduced Fester to the other members of the lab. Fester and Link, the mathematician of the group, hit it off well together, and spent the rest of the afternoon discussing a method of analysis of patterns that Link had been worrying over for the last month.

It was 6:30 when Knowlton finally left the lab that night. He had waited almost eagerly for the end of the day to come — when they would all be gone and he could sit in the quiet rooms, relax, and think it over. "Think what over?" he asked himself. He didn't know. Shortly after 5:00 P.M. they had all gone except Fester, and what followed was almost a duel. Knowlton was annoyed that he was being cheated out of his quiet period, and finally resentfully determined that Fester should leave first.

Fester was sitting at the conference table reading, and Knowlton was sitting at his desk in the little glass-enclosed cubicle he used during the workday when he needed to be undisturbed. Fester had gotten the last year's progress reports out and was studying them carefully. The time dragged. Knowlton doodled on a pad, the tension growing inside him. What the hell did Fester think he was going to find in the reports?

Knowlton finally gave up and they left the lab together. Fester took several of the reports with him to study in the evening. Knowlton asked him if he thought the reports gave a clear picture of the lab's activities.

"They're excellent," Fester answered with obvious sincerity. "They're not only good reports; what they report is damn good, too." Knowlton was surprised at the relief he felt, and grew almost jovial as he said good night.

Driving home, Knowlton felt more optimistic about Fester's presence in the lab. He had never fully understood the analysis Link was attempting. If there was anything wrong with Link's approach,

Fester would probably spot it. "And if I'm any judge," he murmured, "he won't be especially diplomatic about it."

He described Fester to his wife, who was amused by the broad leather belt and the brass buckle.

"It's the kind of belt the Pilgrims must have worn," she laughed.

"I'm not worried about how he holds his pants up," he laughed with her. "I'm afraid he's the kind that just has to make like a genius twice a day. And that can be pretty rough on the group."

Knowlton had been asleep for several hours when he was jarred awake by the telephone ringing. He realized it had rung several times. He swung off the bed muttering about damn fools and telephones. It was Fester. Without any excuses, apparently oblivious of the time, he plunged into an excited recital of how Link's patterning problem could be solved.

Knowlton covered the mouthpiece to answer his wife's stage-whispered, "Who is it?"

"It's the genius," replied Knowlton.

Fester completely ignored the fact that it was 2:00 A.M., and proceeded in a very excited way to start in the middle of an explanation of a completely new approach to certain photon-lab problems he had stumbled on while analyzing past experiments. Knowlton managed to put some enthusiasm in his own voice and stood there, half-asleep and very uncomfortable, listening to Fester talk endlessly about what he had discovered. It was probably not only a new approach, but also an analysis that showed the inherent weakness of the previous experiment, and how experimentation along that line would certainly have been inconclusive.

The following day Knowlton spent the entire morning with Fester and Link, the morning meeting having been called off so that Fester's work of the previous night could be gone over intensively. Fester was very anxious that this be done, and Knowlton was not too unhappy to call the meeting off for reasons of his own.

For the next several days Fester sat in the back office and did nothing but read the progress reports of the work that had been done in the last six months. Knowlton caught himself feeling apprehensive about the reaction Fester might have to some of his work. He was a little surprised at his own feelings. He had always been proud — although he had put on a convincingly modest face — of the way in which new ground in the study of photon measuring devices had been broken by his group. Now he was not so sure, and it seemed to him that Fester might easily show that the line of research they had been following was unsound or even unimaginative.

The next morning, as was the custom in Bob's group, all the members of the lab sat around a conference table. Bob always prided himself on the fact that the work of the lab was guided and evaluated by the group as a whole, and he was fond of repeating that it was not a waste of time to include secretaries in such meetings. Often what started out as a boring recital of fundamental assumptions to a naive listener uncovered new ways of regarding these assumptions, and would not have occurred to the researcher who had long ago accepted them as a necessary basis for his work.

These group meetings also served Bob in another way. He admitted to himself that he would have felt far less secure if he had had to direct the work out of his own mind, so to speak. With the group meeting as the principle of leadership, it was always possible to justify the exploration of blind alleys because of the general educative effect on the team. Fester was there; Lucy and Martha were there; Link was sitting next to Fester, their conversation concerning Link's mathematical study apparently continuing from yesterday. The other members, Bob Davenport, George Thurlow, and Oliver, were waiting quietly.

Knowlton, for reasons he didn't quite understand, proposed for discussion a problem that all of them had spent a great deal of time on previously, with the conclusion that a solution was impossible, that there was no feasible way of treating it in an experimental fashion. When Knowlton proposed the problem, Davenport remarked that there was hardly any use of going over it again, that he was satisfied that there was no way of approaching the problem with the equipment and the physical capacities of the lab.

This statement had the effect of a shot of adrenalin on Fester. He said he would like to know what the problem was in detail, and, walking to the blackboard, began writing out the factors as various members of the group began discussing the problem, and in another column he listed the reasons it had been abandoned.

Very early in the description of the problem it was evident that Fester was going to disagree about the impossibility of attacking it. The group realized this, and finally the descriptive materials and their re-counting of the reasoning that had led to its abandonment dwindled away. Fester began his statement, which, as it proceeded, might well have been prepared the previous night, although Knowlton knew this was impossible. He couldn't help being impressed with the organized and logical way that Fester was presenting ideas that must have occurred to him only a few minutes before.

Fester had some things to say, however, that left Knowlton with a mixture of annoyance, irritation, and, at the same time, a rather smug feeling of superiority over Fester in at least one area. Fester was of the opinion that the way the problem had been analyzed was really typical of group thinking and, with an air of sophistication that made it difficult for a listener to dissent, he proceeded to comment on the American emphasis on team ideas, satirically describing the ways in which they led to a "high level of mediocrity."

During this performance, Knowlton observed that Link stared studiously at the floor, and he was very conscious of George Thurlow's and Bob Davenport's glances toward him at several points of Fester's little speech. Knowlton couldn't help feeling that this was one point at least in which Fester was off on the wrong foot. The whole lab, following Jerry's lead, talked if not practiced the theory of small research teams as the basic organization for effective research. Fester insisted that the problem could be approached and that he would like to study it for a while himself.

Knowlton ended the morning session by remarking that the meetings would continue and that the very fact that a supposedly insoluble experimental problem was now going to get another chance was a further indication of the value of such meetings. Fester immediately remarked that he was not at all averse to meetings for the purpose of informing the group of the progress of its members — that the point he wanted to make was that creative advances were seldom accomplished in such meetings, that they were made by the individual living with the problem closely and continuously, having a sort of personal relationship to it.

Knowlton told Fester he was very glad that Fester had raised these points, and that he was sure the group would profit by re-examining the basis on which they had been operating. Knowlton agreed that individual effort was probably the basis for making the major advances, but that he considered the group meetings useful primarily because of the effect they had on keeping the group together and on helping the weaker members of the group keep up with the ones who were able to advance more easily and quickly in the analysis of problems.

It became clearer, as days went by and meetings continued as they did, that Fester came to enjoy them because of the pattern the meetings assumed. It became typical for Fester to predominate, and it was unquestionably clear that he was more brilliant, better prepared on the various subjects that were germane to the problems being studied, and that he was more capable of going ahead than anyone there. Knowlton grew increasingly disturbed as he realized that his leadership of the group had been, in fact, taken over.

Whenever Fester was mentioned in the occasional meetings with Dr. Jerrold, Knowlton would comment only on the ability and obvious capacity for work that Fester had. Somehow he never felt he could mention his own discomforts, not only because they revealed a weakness on his own part, but also because it was quite clear that Jerrold himself was considerably impressed with Fester's work and with the contacts he had with him outside the photon laboratory.

Knowlton now began to feel that perhaps the intellectual advantages Fester had brought to the group did not quite compensate for what he believed were evidences of a breakdown in the cooperative spirit he had seen in the group before Fester's arrival. More and more of the morning meetings were skipped. Fester's opinions concerning the abilities of others of the group, with the exception of Link, was obviously low. At times, during the morning meetings or in smaller discussions, he had been on the point of rudeness, refusing to pursue an argument when he claimed it was based on the other person's ignorance of the facts involved. His impatience of others led him also to make similar remarks to Dr. Jerrold. Knowlton inferred this from a conversation with Jerrold in which Jerrold asked whether Davenport and Oliver were going to be continued on; and his failure to mention Link led Knowlton to believe that this was the result of private conversations between Fester and Jerrold.

It was not difficult for Knowlton to make a quite convincing case on whether the brilliance of Fester was sufficient recompense for the beginning of this breaking up of the group. He took the opportunity to speak privately with Davenport and with Oliver and it was clear that both of them were uncomfortable because of Fester. Knowlton did not press the discussion beyond the point of hearing them in one way or another say they felt awkward and that it was sometimes difficult for them to understand the arguments he advanced, and often embarrassing to ask him to fill in the background on which his arguments were based. Knowlton did not interview Link in this manner.

About six months after Fester's entrance into the photon lab, a meeting was scheduled in which the sponsors of the research were coming in to get some idea of the work and its progress. It was customary at these meetings for project heads to present the research being conducted in their groups. The members of each group were invited to other meetings held later in the day and open to all, but the special meetings were usually made up only of project heads, the head of the laboratory, and the sponsors.

As the time for the special meeting approached, it seemed to Knowlton that he must avoid the presentation at all cost. His reasons for this were that he could not trust himself to present the ideas and work that Fester had advanced, because of his apprehension as to whether he could give them in sufficient detail and answer the questions about them that might be asked. Yet he did not feel he could ignore these newer lines of work and present only the material he had done or had been started before Fester's arrival. He felt also that it would not be beyond Fester at all, in his blunt and undiplomatic way (if he were present at the meeting, that is), to make comments on his own presentation and reveal the inadequacy Knowlton believed he had. It also seemed quite clear that it would not be easy to keep Fester from attending the meeting, even though he was not on the administrative level that was invited.

Knowlton found an opportunity to speak to Jerrold, and raised the question. He remarked to Jerrold that, with the meetings coming up and with the interest in the work and with the contributions that Fester had been making, he would probably like to come to the meetings, but there was a question of the feelings of the others in the group if Fester alone were invited. Jerrold passed this over very lightly by saying that he did not think the group would fail to understand Fester's rather different position, and that he thought Fester by all means should be invited. Knowlton then immediately said he had thought so, too, and that he thought Fester should present the work, because much of it was work he had done; and that, as Knowlton put it, this would be a nice way to recognize Fester's contributions and to reward him since he was eager to be recognized as a productive member of the lab. Jerrold agreed, and the matter was decided.

Fester's presentation was very successful and in some ways dominated the meeting. He attracted the interest and attention of many of those who had come, and a long discussion followed his presentation. Later in the evening, with the entire laboratory staff present, in the cocktail period before the dinner, a little circle of people formed around Fester. One of them was Jerrold himself, and a lively discussion took place concerning the application of Fester's theory. All of this disturbed Knowlton, and his reaction and behavior were characteristic. He joined the circle, praised Fester to Jerrold and to others, and remarked on the brilliance of the work.

Knowlton, without consulting anyone, began at this time to take some interest in the possibility of a job elsewhere. After a few weeks he found that a new laboratory of considerable size was being organized in a nearby city, and that the kind of training he had would enable him to get a project-head job equivalent to the one he held at the lab, for slightly more money.

He immediately accepted it and notified Jerrold by letter, which he mailed on a Friday night. The letter was quite brief, and Jerrold was stunned. The letter merely said he had found a better position; that there were personal reasons why he did not want to appear at the lab any more; that he would be glad to come back at a later time from where he would be, some 40 miles away, to assist if there was any mixup at all in the past work; that he felt sure Fester could, however, supply any leadership that was required for the group; and that his decision to leave so suddenly was based on some personal problems (he hinted at problems of health in his family, his mother and father). All of this was fictitious, of course. Jerrold took it at face value, but still felt that this was very strange behavior and quite unaccountable, since he had always felt his relationship with Knowlton had been warm and that Knowlton was satisfied, and, as a matter of fact, quite happy and productive.

Jerrold was considerably disturbed, because he

had already decided to place Fester in charge of another project that was going to be set up very soon, and had been wondering how to explain this to Knowlton, in view of the obvious assistance and value Knowlton was getting from Fester and the high regard in which he held him. He had, as a matter of fact, considered the possibility that Knowlton could add to his staff another person with the kind of background and training that had been unique in Fester and had proved so valuable.

Jerrold did not make any attempt to meet Knowlton. In a way, he felt aggrieved about the whole thing. Fester, too, was surprised at the suddenness of Knowlton's departure, and when Jerrold, in talking to him, asked him whether he had reasons to prefer to stay with the photon group instead of the project for the Air Force, which was being or-ganized, he chose the Air Force project and went on to that job the following week. The photon lab was hard hit. The leadership of the lab was given to Link with the understanding that it would be temporary until someone could come in to take over.

Discussion Questions

1 Compare and contrast the value of Knowlton to the laboratory with that of Fester.
2 What could Dr. Jerrold have done to keep Knowlton? Should he have?
3 Formulate advice for Knowlton and Fester. Describe how you would try to persuade either to take your advice.

Two PROBLEM FORMULATION

The difference between an existing state and a desired state can be taken as a general definition for the term *problem*, and it is when he becomes aware of such a difference that an engineer's work begins. The earlier he reaches this awareness the better he is operating, and to the extent he can he should try to anticipate problems and solve them before they have become problems to others.

Polaroid's Edwin Land became aware of the need for self-developing film when his small daughter asked why she could not see a picture of her he had just taken. The rest of the world, in contrast, did not become aware of the need for the Land camera until it became available, and since that time enormous production of the camera has been required to fill that need.

Having become aware of a difference between "as is" and "should be," the engineer immediately faces a trap. His easiest line of thought will be to stop considering the question of how to define the problem he has perceived and instead move on to possible solutions. Implicitly he may be accepting a definition of the problem that excludes better possible solutions than his definition will allow. For instance, a young man who has been directed by his father to cut the grass and finds that chore unappealing could define his problem in a number of different ways. His immediate impulse might be to try to avoid the task by "not hearing" his father, by arguing with him, or possibly by using his allowance to hire someone else to cut the grass. Alternatively, he could define his problem in terms of how to make the task less offensive, perhaps by adapting a motor to the lawnmower or maybe even automating it. Or he could view the problem as one of how to keep the grass short, and seek a way to develop two-inch grass.

Which would be the best problem definition? What problems do automatic ironing machines and permanent-press clothes both solve? Which is a better solution and how might it depend on how the problem is defined? How could anyone form a judgment without considering all the possibilities?

The importance of gathering relevant information about a problem and weighing it carefully to distinguish cause and effect relationships is another facet of problem definition to be observed. An illustration of misdiagnosis is the story of a Japanese pilot during World War II who flew in an attack on an Aleutian island. His plane was hit by groundfire, and he saw his oil-pressure gage reading fall to nil. Believing his engine had been irreparably damaged, he radioed his carrier that he would crash-land on a nearby uninhabited island rather than trying to fly back. For several days thereafter a Japanese submarine surfaced near the island looking for the pilot, but he never appeared.

A few weeks later U.S. troops moved onto the island and found a Japanese Zero aircraft, on its back where it had flipped during crash landing. The pilot hung dead in his harness with a broken neck, but the plane was almost completely undamaged. This was the famous Zero fighter about which the U.S. knew little, except that it totally outclassed fighters the Allies had been able to fly against it. It was brought back and tested by American pilots and engineers who thereby learned its strengths and weaknesses and were able to design a new fighter, the Grumman Hellcat, to surpass it.

In testing the Zero it was found that no groundfire had caused any damage to the engine. However, one bullet had severed the line connecting the oil-pressure gage, causing it to give a false reading. The reader may see several ways that problem definition, both effective and ineffective, figured in this episode.

It is often difficult to determine causes of failure even after the fact, as attested to by the millions of dollars spent by the FAA to determine causes of plane crashes, planes that must be collected scrap by scrap and painstakingly pieced together to reconstruct what caused the crash. Still further millions must then be spent to correct any flaws. This is further illustrated by the recalls by the automobile industry that occur each year.

Even more difficult is the task of anticipating failures before they happen, a procedure that is always desirable and sometimes absolutely essential, as in the space program where there would be no second chance in case of a major malfunction. No formula will give absolute certainty of predicting all possible failures. Basically the engineer must simply be extremely careful about possible weaknesses in his design and try to anticipate as many of the possible things as he can that might go wrong. Two general procedures developed in the space program for anticipating possible failures are fault-tree analysis and failure-mode analysis. The first starts by considering major failure of the total system under consideration, and then asks, "What are all the possible ways a failure like that could occur?" The second type, failure-mode analysis, begins at the component level rather than with the overall system, and asks for each component, "What will happen to the overall system if this component fails?" Both of these approaches amount to simply looking for all the possible things that can go wrong so they can be corrected in advance.

Although no general formula gives *the* answer for how to explain or anticipate failures, there are many formulas that can, at times, help. These include mathematical relationships taught in engineering science courses. Free-body diagrams help in dealing with loads. Laws of conservation for mass and energy provide logical frameworks for delimiting possibilities in situations where something is not working as it should. Application of such physical principles early in the analysis can tremendously simplify the process of looking for possible failure modes by ruling out those that are physically impossible. These are "tricks" that give engineers a substantial edge over the untrained in coping with real or potential failure.

Loss due to failure is one of two kinds of loss that engineers are paid to work against. The second type of loss can be termed "opportunity loss." Stated more positively it refers to opportunities for doing things better and improving upon what exists, even though it is considered acceptable as is. Watches, for example, were an improvement over sundials for keeping time, and for many years the conventional balance wheel was considered an adequate method for measuring time. But then a tuning-fork system was developed that surpassed the balance wheel in accuracy and reliability. Today, tuning-fork watches account for an increasing proportion of the timepieces people prefer to have. Even that is apparently now being surpassed in performance, however, by the all-electronic, quartz-crystal watch, which may be what people will prefer in the future. And beyond that, what further improvements lie ahead? Engineers should answer.

A tendency to look for opportunities to make further improvements is typically more a state of mind left to its natural development than one urged by formal schooling. It is a mental capacity that can be deliberately developed by regular practice, but such practice requires individual initiative. Beyond initial formulation of an engineering problem usually follow further definitions of two types. The first consists of making the general statement more explicit and specific. The pump design called for in the Jack Wireman case that follows was not put simply in terms of the need for a device to pump hydraulic fluid. It was further necessary to specify what kind of fluid, how much of it, at what temperature and pressure, using how much power at what voltage, and with what reliability. Many of these detailed specifications of the design must be quantitative, while others may be in terms of configurations, such as those needed to mate with existing equipment, electrical connectors, and so forth.

It might seem most logical to set forth all such criteria at the outset of the design process and then work toward them until the task is done. Sometimes it is possible to do so when precedent is sufficiently developed in the particular type of design or the new design is patterned very closely after a similar design. Often, however, there is sufficient newness that what the criterion should be is itself somewhat uncertain, and as the design proceeds it becomes necessary to modify the design targets. Cost overruns are often one result of such modification. Sometimes these arise out of poor performance on the part of the designing company, and at other times as a result of changes in the initial design requested by the customer.

Examples of changing design criteria have become familiar in the auto industry, where new standards have been imposed by the government to reduce polluting emissions and to increase safety. At times the uncertainty of the criteria have been emphasized by public argument between government officials urging tighter restrictions and auto executives who claim they are not yet technologically possible and cannot be developed in time to meet the standards. When criteria truly do turn out to be impossible to meet, they are usually relaxed in some way, and at other times unexpected breakthroughs make it possible to raise them still further.

Some of the changes in criteria can arise when a second type of further definition takes place, namely the breaking down of the initial overall design problem into subproblems, and then the further breaking down of these subproblems. For instance, the initial problem statement in the Jack Wireman case may have been to design an airliner to meet certain performance criteria at a certain price. As the airplane design was refined, the need for a certain hydraulic pump was found, leading to specifications for the pump. Design of the pump to meet these requirements has in turn led to further subproblems, as shown in the case. Whether these are viewed as further changes in specifications or as new subproblems is less important than the point that the nature of problems tends to unfold as engineering progresses. It is not always possible to define problems at the outset of a design. Specifications must at times be changed by developments that reveal either new needs or new opportunities, or both.

Yet it is important to be as specific as possible when asking for or undertaking a design, and to anticipate subproblems so as to reduce the number that will come as unpleasant surprises. Expensive backtracking may never be eliminated, and understanding among different people concerned with the design may never be perfect, but care in problem definition is basic to a positive beginning.

Exercises

1   Considering the millions of dollars spent on recalls of new automobiles to correct defects, what policies would you recommend be adopted by Detroit auto makers: (1) at the top executive level? (2) at the engineering department level? How would you go about putting such policies into effect if you worked at either level?

2   Describe some physical object that in your personal experience failed to perform as it should. Explain what physical laws apply, what caused the failure, and how the design might be modified to prevent further failures. Do you recommend that such changes be incorporated? Please explain.

3   "It is sometimes impossible for an engineer to solve one problem without creating others." Discuss the validity of this statement giving examples from your experience.

4   "Air travel today is unsatisfactory." What is the problem? How many ways can it be defined and how does the choice of definition delimit possible solutions?

5   Develop one or more statements of the problems for which each of the following is a solution:

a   Keep-off-the-grass sign
b   Automobile
c   Newspaper
d   Sunglasses
e   Tape recorder
f   I-beam
g   Slide rule
h   Digital computer
i   Analog computer
j   Shock absorber

6   Think of three physical things you would like to have, plus three you think others besides yourself would like to have, that have not yet been developed. Write an assignment for someone else to design specifically those six things.

7   Write general specifications for:
a   a transportation system in the vicinity of your campus.
b   a parking system for the vicinity of your campus.

8   Discuss what might be involved in developing more detailed specifications in the above two systems, and what sorts of subproblems might be expected.

9   Suggest some detailed specifications that might apply to a familiar implement, such as a toaster or a bicycle. Consider them from the point of view of a shop charged with fabrication, a salesman for the producer, and a customer. How might these three viewpoints contrast?

10   For one of the cases that follow in this chapter choose some facet of the design or one subproblem and list one or more physical laws that apply, telling how they delimit the problem. Illustrate how the law might be analytically applied (with a free-body diagram or energy balance, etc.).

TERRY ABBEY   Deep-well Waste Disposal

If the casing should burst on the new well for disposing of waste acid deep underground, it could pollute fresh water supplies. How adequate is the present design?

"Our hunch is that the deep well over at Great Northern Steel collapsed from hydrostatic pressure in the annulus when the flow was shut off. We think our design will prevent that, but we want to be as sure as we can. We also want to be sure that we will spot any irregularities in operation early." Bob Martin, superintendent of Support Facilities Engineering at the Northern Indiana plant of Central Steel Corporation was referring to a new 4,300-ft well being drilled for disposal of "waste pickle liquor," a mixture primarily of hydrochloric and sulfuric acid used to take chemical scales off steel ingots before they were rolled into ingot. This acid (specific gravity ranging from 1.1 to 1.2), which was at present being pumped into a nearby river, would instead pass down a 2⅞-in. diameter, Fiberglas-reinforced plastic pipe to displace salt water in a sandstone layer' running from 2,200-ft-deep down to 4,300 ft.

Terry Abbey, who had been with Central for two years since graduating in civil engineering from a prominent Michigan university, had taken over as engineer in charge of the waste-acid disposal project in November, 1966. It was now February, 1967, and detail design of filtering and pumping facilities was in progress. Drilling was scheduled to start in May. One part of the design not yet fully decided was what instrumentation should be provided on the well. Terry thought he should review this question, particularly as it concerned possibilities of leakage or collapse of the pipe. It was expected that the well of Terry's company, Central Steel, would in principle be similar to that of the well already sunk by Great Northern, which had recently collapsed twice. The major exception was that special precautions against such a failure would be taken. A discussion of the waste disposal problem, which appeared in *Business Week*, is reprinted in Exhibit 1 — Terry Abbey.

Existing Well at Great Northern

Great Northern first considered such a well in 1959, received approval for drilling from the state of Indiana in August, 1964, started drilling on September 14, 1964, and "bottomed out" on October 4, 1964. As shown in Exhibit 2 — Terry Abbey, the well provided essentially two concentric tubes, an inner pipe of 2- to 3-in. diameter called the injection tube, through which acid was pumped down, and an outer annulus (casing) through which water was pumped. The water outside the plastic pipe served to counterbalance the pressure of the acid standing inside, thereby supporting the reinforced plastic injection tube against bursting and also providing a barrier to the acid against corroding the outer casing if the plastic pipe failed. At the bottom the annulus was partially sealed by a packer, which had holes in it to allow a trickle of water (about 1 gpm) to flow through. This trickle of water was to keep acid away from the bottom of the well and to prevent upward migration of the acid.

The well casing, after passing down through several horizontal layers of shale that served to isolate the acid permanently from the surface, ended in porous Mt. Simon sandstone presently saturated with salt brine (three times as saline as seawater) at a temperature of 80°F. The natural pressure was enough to force this brine up to within 400 ft of the surface through any hole drilled in the shale. Consequently, the acid had to be pumped down to force this brine back through the sandstone at the desired flowrate. First, however, 35 million gal of fresh water were pumped into the sandstone so the acid could not mix with the salt water and form precipitates that might plug the sandstone (12 percent porosity). The acid was to pass through a 0.6-micron filter before entering the well so that no particles could plug the sandstone. Total costs of the well plus ancillary filtering and pumping facilities were estimated at around $2 million for installation and the first seven years' operation. Disposal of acid through the Great Northern well began April 3, 1965.

Two failures followed shortly thereafter during temporary shutdowns of the Great Northern pumping station. Both times the inner reinforced plastic pipe (injection tube) carrying acid collapsed inward at a depth of 550 ft. To cure the problem, Great Northern installed a section of 2-in. diameter pipe in place of the original 3-in. pipe for 400 ft above and 400 ft below the 550-ft mark to provide greater strength against crushing. By early 1967 there had been no additional failures. A cost comparison estimated for the well system as against the alternative of chemical neutralization of the acid appears in Exhibit 3 — Terry Abbey.

Central Steel's Design

The proposed new well for Central, being 100 yd inside of the Central Steel property line and only a few miles from Great Northern's well, was expected to have essentially the same geological environment. Its capacity was initially to be 120 gal of acid per day at 150°F, possibly increasing ultimately to 300 gal as steel production was expanded. Provision would be made to store 100,000 gal of acid in each of two tanks near the well and to pump river water (heated to a minimum of 50°F in the winter to prevent precipitation) in place of acid when the flow of acid from the plant stopped. Central engaged a well consultant and the same drilling company that had sunk Great Northern's well.

Several alternatives had been suggested as ways to preclude a collapse similar to the one that had occurred at Great Northern. One was to use 2⅞-in. tubing, and in other respects make the Central well like Great Northern's final design (cross-sectional dimensions appear in Exhibit 4 — Terry Abbey). This approach was favored by the driller. The consulting geologist, however, recommended that Central's well not use a packer at the bottom between the inner and outer casings. He pointed out that by eliminating the packer, water would not be restricted in flowing down the annulus, so that pressures of water and acid would remain equal at the bottom of the injection tube. The packer at Great Northern, he said, had probably held the water column back during shutdown as the acid-level dropped, thereby imposing a high differential hydrostatic head in the annulus, which crushed the inner pipe. Furthermore, the packer was difficult to install, particularly to allow for slip of the inner pipe needed for differential expansion due to temperature changes in the flowing fluids. (The coefficient of linear expansion for the plastic pipe was 9.5×10^{-6} in./in./°F.) On the other hand, the driller pointed out that the packer might help prevent acid from migrating back up the annulus, and would serve to hold the injection tube from falling in case of a break.

Another alternative considered was to plug the annulus at the bottom and fill it solid with plastic. "The

plastic might cost $10 to $15 a gallon," Terry observed. "I don't know whether there would be problems with thermal expansion, but if so the plastic might crack and let the acid eat through the steel. Then it could flow out into some fresh-water zone without our even finding out until it was too late. Also, we wouldn't be able to change capacity without drilling out all the plastic. Ultimately, we think, we may want to put in a 4½-in. diameter injection tube (center pipe for acid). We could maybe have used that size now, but management told us to stay pretty close to Great Northern's design, so we chose not to go to a larger diameter."

The final decision was that Central would use a 2⅞-in. pipe all the way. Whereas Great Northern had used centrifugal pumps on both annulus and injection tube, however, Central would use a positive displacement pump on the annulus and centrifugal on the injection tube.

Terry also did some checking to see if stronger pipe were available, but did not find any. He wrote the Poxy Pipe Company, from which the injection tube material was bought, to obtain figures for allowable stress on the pipe, and received the figures shown in Exhibit 5 — Terry Abbey. Shown in Exhibit 6 — Terry Abbey are pressures required to force varying flowrates of acid into the sandstone formation, and head loss due to friction of the pipe for various flowrates. Terry obtained these figures from the consultant.

"Deciding what information you need and collecting it is 90 percent of a job like this," Terry observed. "In school you may get the impression that engineering work involves lots of formulas and calculations, but here that turns out to be just a small part. You're on the phone, you're writing letters and specifications, reviewing catalog data, and buying things through the purchasing department."

The idea for this project had initially been suggested in a brief memo called an engineering request, which had come from an assistant to one of the vice presidents. A "project scope" was then prepared, outlining the proposed design and estimating the work schedule and costs. Approval of an engineering committee and of the board of directors was then obtained, as was required on all engineering projects, after which preparation of detailed specifications, solicitation of contractor bids, and execution of the project ensued. Practically all new equipment for the plant was purchased through specifications, and it was the project engineer's responsibility to assure that they were appropriate, complete, and complied with by suppliers.

One of the senior engineering managers at Central Steel observed that the engineering work done by steel companies does not normally include the design of steelmaking machinery. "We get ideas for machinery improvement or for new equipment that we need and pass these along to companies like Bliss and

Blaw-Knox, and they do the engineering and manufacture the machines. Our engineers write the specifications by which we then buy the equipment. This involves a lot of technical variety, and it's why we look for problem-solving ability first when we hire engineers, not for specialized degrees. Sometimes a superintendent of the plant will call for a crane when the job could be done with a jack. Our engineer has to have the horse sense to recognize something like this and also enough diplomacy to make the superintendent feel good about being set straight."

Projects with which Terry had worked since joining Central had included such things as choosing a sump pump for a degreaser pit, selecting and ordering an automatic sampling device for sewer outfalls, and preparing purchase and installation specifications for a pump and related controls to remove waste from a coke plant. He had entered the acid-disposal project at the stage of preparing specifications for the system and within 3 months had been made engineer in charge of the project. "Every day you learn something new in a job like this," Terry commented, "especially from vendors. I learn specifics about instrumentation from them a lot faster than from catalogs. The basic understanding, of course, you have to get from school or textbooks so you know what you're looking for and what questions to ask."

Precautions

"One question I think I should check on here is what pressures act on the injection tube for different flow-rates and acid densities," he continued, "and the calculations shouldn't be too hard to make. What, for instance, is the worst situation that could arise? And can the Great Northern failures really be explained on the basis of hydrostatic pressure, or was it something else like 'waterhammer,' as some others believe? Bob Martin has been asking what might happen if gas bubbles go down the acid line and float back up the annulus, which he thinks is also a real possibility.

"Then there is the question of what safety valves and alarms we should specify for the system. To order instruments and controls we'll have to know scale ranges of the things like pressure and flows we want to measure and regulate. Hopefully, nothing will go wrong. But if it does, we'd better catch it before it leads to more trouble."

Discussion Questions

1 Define the problem faced by Terry Abbey as of the end of this case at least three different ways.
2 Propose a definition of the problem which Terry's company faced and which led to the well project.
3 Describe how you would attack the problem as of the end of the case. Carry out your approach as far as you can and state your conclusions. State the degree to which you are uncertain about them and what you would do about it if you were really there.

Exhibit 1 — Terry Abbey Watery grave for Lake Michigan? *

Pollution-control officials worry that Lake Michigan may soon become a ghost lake — chiefly as a result of industrial waste. But some companies are striving to reverse the tide.

Lake Michigan means different things to different people. Bathers in four states love its 700 miles of shoreline, boaters love its great waves and oceanic distances, Chicagoans revere it as their source of drinking water.

But for everyone in the area, the 22,400-sq.-mi body of water has a darker side — as a convenient receptacle for industrial waste and municipal sewage, it is becoming a colossus of pollution.

Pollution-control agencies fear that in 10 years, despite present abatement programs, Lake Michigan will become another Lake Erie — clogged with algae and murky with waste. The comparison is a chilling one, for if Lake Erie is not already dead, it is moribund. Commercial fishing has nearly disappeared, and at its dead core is a 2,600-sq.-mi. patch of water devoid of all oxygen in summer. "Grossly polluted, aesthetically displeasing, and even hazardous to health" is how George Harlow, director of the Federal Water Pollution Control Administration in Cleveland, describes the water around the major Erie ports.

The big offender. If Lake Michigan is not yet the same sort of ghost lake, it soon could be. Of prime concern to pollution-control authorities is the vast industrial complex — steel, oil-refining, and chemical companies particularly — along the lake's southern basin, which stretches about 25 miles from southeast Chicago through Gary, Ind. One official estimates that "70% of Lake Michigan's pollution is generated in this small area." Industry is trying to make amends. Some 20 companies operating along the Indiana portion of the basin will probably spend upwards of $100 million by 1970 to meet that state's stringent new control requirements.

Yet industry alone is not responsible. Representatives of the U.S. Army Corps of Engineers were summoned to an informal Congressional hearing by Rep. John C. Kluczynski (D-Ill.) early this month to explain why the engineers continue to dump contaminated dredgings from the Indiana Harbor Canal into Lake Michigan. (The engineers clean the canal every 18 months and the dredgings are prodigious: The canal is bordered by Inland Steel Co. and Youngstown Sheet & Tube Co.)

Switch. Corps officials replied that because the cost of landfill disposal methods was still prohibitive, dumping would continue through at least 1970. But some days later — in an apparent response to public pressure — the engineers announced they had agreed with Inland and Youngstown to deposit this year's remaining dredgings — 120,000 cubic yards — in diked landfills at both companies.

Col. Edward A. Bennett, chief engineer for the Corps' Chicago district, concedes that critics have a legitimate case to make. But his rejoinder is of small consolation to fresh-water enthusiasts: "Since Indiana Harbor flows into the lake anyway, we're simply moving the dredgings from one point to another, within the system. We're really not adding to the problem."

Such answers are all too typical of the "Who, me?" attitude of many companies along the lake's southern basin — at least in years past. Or so says one official of the Federal Water Pollution Control Administration. The same official points out just why Lake Michigan's problems are so acute. "The recovery capabilities of a fresh water lake aren't nearly as good as those of, say, a river," he explains. "A river is constantly flushing out, but Lake Michigan just sits there, almost like a giant bathtub."

* Reprinted from the Oct. 21, 1967 issue of *Business Week* by special permission. Copyright, 1972 by McGraw-Hill, Inc.

Dying lake. FWPCA, the water pollution control arm of the Dept. of the Interior, and Indiana and Illinois state agencies are working together, but they wonder aloud whether present regulations are not a case of too little and too late.

Until late 1964, regulation of water pollution in the area was flimsy at best. In December of that year, the federal government called three agencies — the Chicago Sanitary District, the Illinois Sanitary Water Board, and the Indiana Stream Pollution Control Board — to the conference table. At their meeting the following March, the conferees established a technical committee to develop water quality criteria for the southern end of Lake Michigan.

Option. Soon after the conference, Congress passed the Water Quality Act of 1965, giving states the option of developing sets of water standards and submitting them to the federal government for approval. Using criteria developed from the 1965 conference as a base, both Indiana and Illinois offered programs, complete with timetables for the installation of pollution abatement facilities. Indiana has been cleared and Illinois should be shortly.

The Indiana enforcement plan mentions companies by name. Among them are three steelmakers — Inland, Youngstown, and U.S. Steel Corp. — considered by government pollution-control experts to be the biggest polluters of all. The companies expel large quantities of ammonia, cyanide, phenol, iron oxides, oils, and sulfuric acid into waterways flowing into Lake Michigan. Says a bitter Robert J. Bowden, chief of FWPCA's Calumet area surveillance project: "The chemical and oil industries themselves would cause serious pollution problems. But with the steel companies there, you hardly notice the others."

Reluctant. Inland, Youngstown, and U.S. Steel use close to 3-billion gallons of Lake Michigan's water each day. All three are reluctant to reveal the quantities of pollutants they discharge into the lake, but FWPCA cites what it considers a fairly typical example: 37,000 gallons of oil each day pour into tributaries of Lake Michigan.

Steelmakers in Indiana say they can put pollutant-abatement facilities into operation on schedule. Inland will sink $17 million into its new program to meet the standards, Youngstown about $12 million. Currently, the three steelmakers have until June 30, 1970, to get started. Because of the volume and complexity of their effluent, they have 18 months more than other companies on the southern basin of the lake. But eagerness to get pollution control under way may wipe out this cushion. Insiders report that Indiana's Stream Pollution Control Board will meet with the steelmakers next week to discuss a Dec. 31, 1968 deadline, the same as Illinois.

Later, please. The steel companies would prefer getting state and local agencies on a 1970 schedule. John Brough, director, Air and Water Control at Inland (companies avoid the word pollution in titles), explains why steelmakers have waited so long in stepping up their control efforts: "Since 1950, all our new facilities have included abatement. Before that, facilities were controlled to meet specific problems. Most of these were controls to pull out bulk materials." Says another steel official: "We've known the stuff we dump is deadly, but we've been doing the best we can. We don't see any evidence that other industries are working as hard."

The industry's defense of its abatement program is this: We'll do all we can, but we are accountable to our stockholders, and pollution control equipment doesn't put a cent in their pockets.

The effluent society. But the widening concern over pollution, even from stockholders, is making this defense untenable. U.S. Steel doesn't expect to stop pollution-control efforts and expenditures when it has met present deadlines, be they 1968 or 1970. Charles A. Bishop, director of U.S. Steel's abatement program, looks forward to more, and tougher, standards and isn't griping about finding solutions.

U.S. Steel's South Works and Gary mills, hugging the Lake Michigan shoreline, already have made such improvements as separate sanitary sewage systems.

Indirect. It is true that the steel-making process produces very little phosphate waste, a key ingredient of nutrients fostering algae growth. But it does produce ammonia, which contains nitrogen, another contributor to nutrients.

What authorities fear most is the increasing cost of treating water for drinking. FWPCA's Bowden points out that water treatment costs decrease as one moves farther away from the southern basin area. He noted that Hammond, Ind., which is in the heart of the area, pays about $36 per million gallons to have its water treated. The cost in Chicago's central district, however, is only half that. "If pollution continues at this pace," says Bowden, "water could become as expensive as your food bill. Would you take a shower in the morning if it cost you as much as your breakfast?"

Exhibit 2 — Terry Abbey

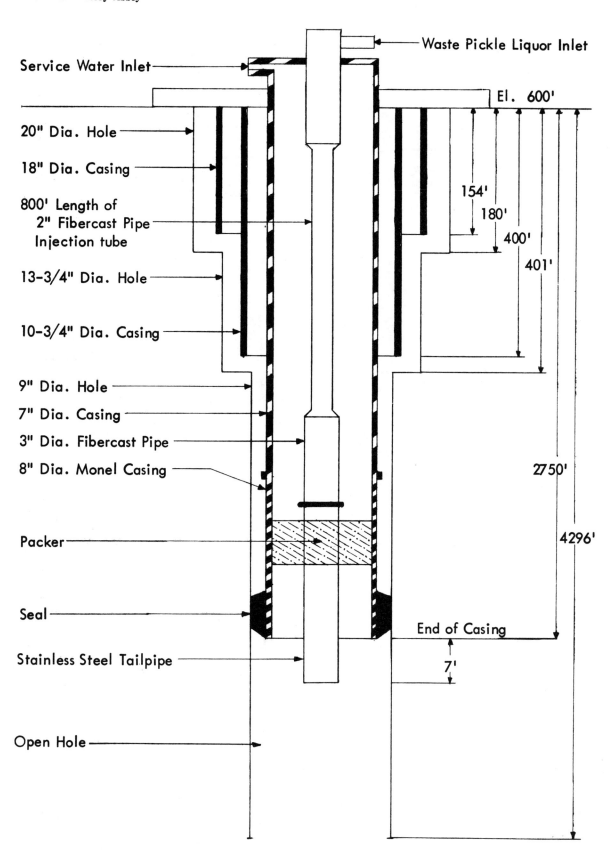

Waste Pickle Liquor Inlet

Service Water Inlet

El. 600'

20" Dia. Hole

18" Dia. Casing

800' Length of
2" Fibercast Pipe
Injection tube

154'

180'

400'

13-3/4" Dia. Hole

401'

10-3/4" Dia. Casing

9" Dia. Hole

7" Dia. Casing

3" Dia. Fibercast Pipe

8" Dia. Monel Casing

2750'

4296'

Packer

Seal

End of Casing

Stainless Steel Tailpipe

7'

Open Hole

Exhibit 3 — Terry Abbey   Great Northern waste-acid disposal cost comparison between lime neutralization and deep-well methods of waste disposal

	DOLLARS PER MONTH	
	Neutralization	*Deep Well*
Labor (including benefits)	3,214	875
Repair and maintenance	10,807	2,312
Electricity	576	200
Steam; Utilities	4,014	0
Water	64	400
*Lime	9,295	0
Supplies	910	197
Services	832	291
Technical and administrative	3,260	907
TOTAL	32,972	5,182

* Cost of lime delivered has increased 29.9% since the above lime cost was calculated.

Exhibit 4 — Terry Abbey

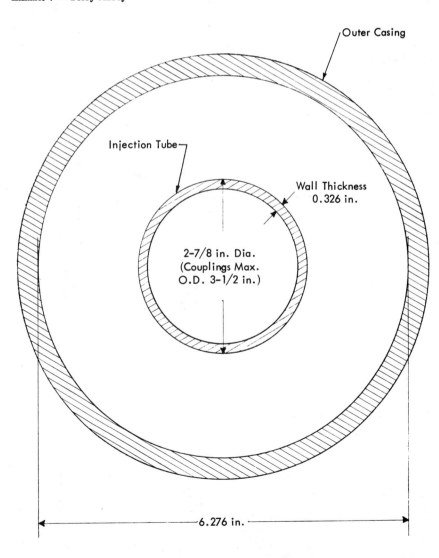

Exhibit 5 — Terry Abbey Fibercast pipe

February 23, 1967

Specifications for 2⅞" O.D. tubing

	OPERATING	ULTIMATE
1. Internal Pressure		
150° and 200° F.	575	2,000
300° F.	250	1,500
2. Collapse		
150° F.	500	2,500
200° F.	375	1,875
300° F.	250	1,250
3. Tensile (thread shear controlling)		
150° and 200° F.	7,500	40,000
300° F.	4,000	30,000
4. Coupling Dimensions		
Same as Catalog — Figure 1 × 2½"		

Exhibit 6 — Terry Abbey Consultant's estimates of pressures required for operation

A. Feet of Pressure Required to Force Annulus Water (at 1 gallon/min.) and Waste Pickle Liquor (Specific Gravity of 1.2) into the Sandstone Layer at Various Flowrates.

ACID FLOWRATE (GAL./MIN.)	FEET OF PRESSURE HEAD REQUIRED (FT.)	
	Water	Waste Pickle Liquor
0	2410	2010
50	2475	2060
100	2545	2120
150	2615	2180
200	2680	2230
250	2750	2290
300	2820	2350

B. Feet of Pressure Required to Force Waste Pickle Liquor (Specific Gravity of 1.2) through 2,500 Feet of Tubing of Various Sizes Against Wall Friction of the Tubing.

ACID FLOWRATE (GAL./MIN.)	REQUIRED PRESSURE HEAD (FT.)			
	Injection Tubing O.D. (in.)			
	2⅜	2⅞	3½	4½
50	188	50	18	5
100	650	175	63	19
150	1375	350	127	40
200	2380	625	225	68
250	3500	900	325	100
300	5000	1250	450	140

JACK WIREMAN

Part 1

Failure of a Ball Bearing

A bearing has failed prematurely on an electric motor sent to the customer for acceptance testing. Do catalog calculations say it was adequate for the given load? What is the company engineer's responsibility?

In September, 1963, Jack Wireman, a mechanical engineer at Task Corporation, was faced with the question of what should be done about the failure of an electric-motor shaft bearing. The motor was a new design to power an aircraft hydraulic pump. A contract for 300 of the motors had been placed with Task, and the terms of this contract required that the motor be capable of operating for 2,500 hr without failure. When the motor in test began to draw excessive current after only 1,800 hr of operation, it was shut down and disassembled. Examination showed severe pitting and wear of the balls and races in the front shaft bearing. Pictures of the bearing after failure appear in Exhibit 1 — Jack Wireman (Part 1). A cross-section drawing of the motor and hydraulic pump to which it attached appears in Exhibit 2 — Jack Wireman (Part 1).

Task was at this time a company of 140 employees, located in Anaheim, California, with sales of around $2 million per year. The company had been growing but was nonetheless still trying to break even after several years of serious problems, which had left it in a somewhat precarious financial position. Products manufactured by the company included electric motors, pumps, fans, blowers, refrigeration systems, and measuring instruments for wind-tunnel testing. The electric motors ranged in size from fractions to hundreds of horsepower, and were for special applications where high performance such as great speed, low weight, or small volume per horsepower were required. Confidence in Task's ability to design high-performance motors was reflected in a comment by Elmer Ward, the company's president: "I once told a man that he could have anyone he wanted design him an electric motor, and if we couldn't come up with a better one in 24 hr, I'd pay $100. He never collected, and the offer is still open."

In April, 1962, Task was invited by the Geyser Pump Company to bid on the job of designing and producing an electric motor to be used for driving a positive-displacement, aircraft hydraulic pump of the piston and wobble plate type. Geyser was, in turn, preparing its own bid to make the pump, attach the motor to the pump, and sell the completed units to the Thunder Aircraft Company for installation in Thunder 99 aircraft.

Two such units were required per aircraft. Each pump would operate continuously during flight. A specification described the pump as "the primary source for the flight power-control units as well as an alternate source of hydraulic pressure. It is also used as a hydraulic pressure source for use during preflight testing and routine aircraft maintenance operations." In addition to these electrically powered hydraulic pumps, each Thunder 99 had its main pumps, which were driven mechanically by the turbojet engines. Operating requirements of the electric pump units included the following:

Hydraulic Fluid	Skydrol 500A [See fluid specification, Exhibit 3 — Jack Wireman (Part 1)]
Fluid Temperature	65°F to 180 ± 5°F
Inlet Pressure	45 to 50 psi
Outlet Pressure	3,000 ± 50 psi
Flowrate	1 to 6 ± 0.1 gpm
Input Power	400 cycle alternating current at 200 V
Output Power (motor only)	11.75 hp
Weight (motor only)	18.75 lb
Efficiency (motor only)	65 percent
Maximum Motor Case Temp.	225°F
Explosion-Proof Life	2,500 hr with "no failure of parts or excessive wear"

In addition to these operating conditions, the motor was required to comply with eight U.S. Government specification documents, eleven nongovernment specification documents, and various other requirements relating to torque at different speeds, rotational acceleration at 65°F, flammability, corrosion and fungus resistance, electrical bonding, radio interference, noise level, fume propagation, safety wiring, dielectric arc-over resistance, and fluid leakage under pressure.

Testing of the performance and life was to be done by Geyser Pump on two qualification units. A schematic diagram of the $1,000 test rig, which cost around $1,007, is shown in Exhibit 4 — Jack Wireman (Part 1). In the life test, each motor was to be run through several different cycles as described in the schedule of Exhibit 5 — Jack Wireman (Part 1).

After some discussion among engineers of Task and Geyser, it was decided that a 6,000-rpm synchronous unit would allow the best compromises of weight and size. Task engineers planned on the motor being liquid-cooled by the hydraulic fluid enroute to the pump. The fluid would, after passing through a 10-micron filter, enter one side of the motor at the front, flow back through annular slots around the stator, circulate forward through the rotor, then pass through a centrifugal impeller at the front of the motor and into the Geyser pump for boost up to 3,000 psi. Both front and rear bearings of the motor would be kept fully immersed by the hydraulic fluid as it circulated through to the pump under a pressure within the motor of around 50 psi. A schematic of this flow pattern appears in Exhibit 6 — Jack Wireman (Part 1). Task offered to design and build the motors at a price between $600 and $1,000 each, depending on the quantity ordered.

Geyser Pump Company won, a contract for the motor-pump units from Thunder and gave Task a contract to provide 300 motors in total over a 12-month period. The first two units would be for qualification tests to assure that the motors performed as required. All remaining units would be for installation in Thunder 99 aircraft. It was expected that contracts for more of the motors would be placed with Task in the future since Task was the only company selected to make them. If, however, the motors did not meet the performance specifications of the contract, Geyser could refuse to accept them from Task and would not have to pay for them. Similarly, by the contract between Thunder and Geyser, Thunder could refuse to accept the complete pump units if either the motor or the pump did not measure up to specified performance.

When the Task proposal was accepted, engineers at Task proceeded with detailed design of the motor. Electrical components, numbers of turns, amount of iron, and other aspects of the electromagnetic circuitry were determined. Dimensions of the path for circulating the hydraulic fluid and the shape of the

centrifugal impeller were designed. The shape and dimensions of the aluminum motor-housing had to be arranged to fit the electrical and hydraulic requirements and also to mate with the Geyser pump. Power transmission between motor and pump was to be through a spline, the dimensions of which were given by Geyser. Shaft dimensions of the motor were picked by Task engineers to transmit expected loads with abundant safety and to blend with the required spline.

Three types of loading were expected on the motor shaft: torsional, radial, and axial. Torsional loading would be primarily due to the torque output of the motor. Radial loading would be mainly caused by the weight of the rotor and impeller (5.3 lb), and by rotor dynamic imbalance, which by careful manufacture, testing, and balancing would be limited to 5 lb. The main axial load on the motor shaft was expected to be imposed by the pump shaft, which was axially preloaded by a spring in the pump. The pump had a male spline that was pushed into the mating motor shaft, compressing the spring, according to Geyser engineers, to a force of 50 to 70 lb.

Ball bearings made by the Orbit Corporation were picked for the shaft. The use of Orbit bearings had become something of a standard practice at Task after problems had been encountered with bearings of several other manufacturers. Orbit specialized in bearings of high precision. Although relatively expensive, such bearings were considered highly reliable.

The front bearing chosen for the motor was an Orbit No. 204SST5 with an outer-race centered * phenolic retainer and a contact angle† of 14°. This bearing was chosen because it apparently provided a sufficient margin of load capacity and its inner race was large enough to slip easily over the internally splined shaft, and its outer race would fit conveniently into the motor housing. Radial allowances between 0.0001 and 0.0003 in. were provided between inner race and shaft and between outer race and housing. To absorb axial loading on the shaft, the inner race of this bearing was rigidly fixed to the

* "Outer-race centered" refers to the manner in which the ball cage or retainer of the bearing is centered. In most commonly used bearings, the ball case is centered by contact against the inner race. At high rotational speeds, centrifugal force imparted by the cage can make it difficult for lubricant to reach the inner race. Also, at high speeds the cage may become centrifugally expanded and thereby lose centering contact. By centering through contact with the outer race, such loss of centering contact is better controlled for many applications.
† "Contact angle" refers to the angle between a line perpendicular to the shaft axis of rotation and a line through both the center of a ball and the point at which that ball contacts with the outer race. The initial contact angle is defined by assuming contact with no deflections and is fixed by radial looseness.

shaft by a nut and lockwasher, and the outer race was fastened into the aluminum motor housing by pieces of the housing being bolted together.

A small bearing, Orbit No. 203SS5, was picked for the opposite end of the shaft. This rear bearing was not rigidly fixed to either shaft or housing, so it could move axially to allow for dimensional variations of manufacture and for expansion and contraction of motor parts with changes in temperature. A radial allowance of 0.0004 up to 0.001 in. was left between outer race and housing, and an allowance of 0 up to 0.0003 in. between inner race and shaft on the rear bearing. To ensure proper bearing geometry and to prevent the bearing from having its outer race turn in the housing, a pair of belleville springs were compressed with a force of 4 to 26 lb between the rear of the housing and the outer race.

Although no detailed life calculations were made on the bearings, design experience indicated to Task engineers that the bearings would last well beyond the 2,500-hr required life, and that if failure occurred, it would first happen elsewhere in the motor. It was assumed that the greatest loads on the front bearing would be axial and less than 100 lb, since the maximum pump preload spring force predicted by Geyser was 70 lb. The Orbit 204, it was believed, would be more than adequate for such loading. For the tail of the shaft, a smaller bearing was considered sufficient. There would be no need for the rear bearing to be large enough to slip over the spline, and the axial loading on it was expected to be very small, equivalent to the belleville-spring load. Both bearings cost around $10, but by using the smaller bearing at the rear, there would be a few cents saved on each motor. Another benefit of the smaller bearing would be lighter weight. Excerpts from an Orbit Company catalog giving specifications on the chosen bearings appear in Exhibit 7 — Jack Wireman (Part 1).

The first qualification test motor was made in the Task shop and tested about 3 hr on the Task dynamometer, without symptoms of difficulty. After this test, the motor was shipped, in late April, 1963,

to the Geyser Pump Company where it was mated to a pump, installed in the test rig, and started on life tests according to the schedule shown in Exhibit 5 — Jack Wireman (Part 1). After 1,800 hr, Geyser technicians monitoring the test noticed that the motor was drawing excessive current. They shut the motor down and removed it from the pump. Turning the shaft over by hand, they could feel roughness in the bearings. (It wasn't possible to hear the bearings during the test because the pump made too much noise.)

At 10:00 A.M. the next day, several Geyser engineers appeared at Task with the motor asking for an answer to the problem. The motor was immediately disassembled in the Task shop. The front bearing had become severely galled, and the grooves of the races had been widened by wear in the same direction as the preload — about 0.040 in. — all the way around. Wear was so extensive on both balls and races that the engineers could see no clues in the bearing as to why it had failed.

Meanwhile, the second qualification test motor had been finished and was ready for testing. Tooling had been completed at a cost of over $100,000 and production was under way, with over a dozen motors complete and others in various stages of manufacture. Geyser Pump was progressing similarly in production of pumps to match the motors. Task top management asked Jack Wireman, a mechanical engineer, to prescribe action.

Discussion Questions

1 Define Jack Wireman's problem at least three different ways. Does it matter how he defines it? Why or why not?
2 Describe how you would approach the solution to this problem. What steps would you take and why?
3 Carry out those steps of the solution you are able to, and prescribe action for Mr. Wireman.

Exhibit 1 — Jack Wireman (Part 1)

Exhibit 2 — Jack Wireman (Part 1)

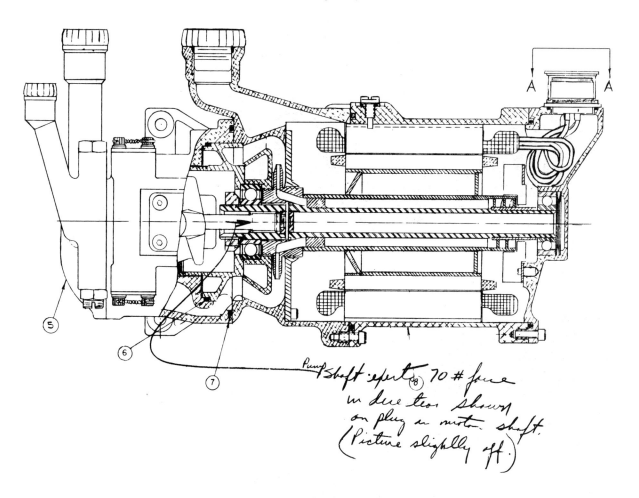

Pump Shaft ·exerts 70 # force
in direction shown,
on plug in motor shaft.
(Picture slightly off.)

Exhibit 3 — Jack Wireman (Part 1) Physical and chemical properties of Skydrol (catalog excerpts

PROPERTY	SKYDROL 500A	SKYDROL 7000
Appearance	Clear, purple liquid	Clear, green fluid
Odor	Mild, pleasant	Mild, pleasant
Autogenous ignition Temperature, (ASTM D-286-58T)	Above 1100°F. (593°C.)	1060°F. (571°C.)
Pour Point, (ASTM D97)	Below −85°F. (−65°C.)	Below −70°F. (−57°C.)
Neutralization number, (ASTM D-974-58T) mg. KOH/gm.	0.01	0.01
Specific gravity, (ASTM D941) 77°F. (25°C.)	1.065	1.086
Viscosity index	+238	+160
Viscosity, CS, (ASTM D445)		
210°F. (99°C.)	3.95	4.00
100°F. (38°C.)	11.70	15.50
Pressure Viscosity, CS		
100°F., @ 4000 psi	20.8	Not available
Thermal conductivity, 82°F. (28°C.)	0.0777 Btu/Hr./Ft.^{2} °F./Ft. (32.2×10^{-5} Cal/Sec. CM.^{2} °C. CM)	0.0723 Btu/Hr./Ft.^{2} °F./Ft. (29.9×10^{-5} Cal/Sec. CM.^{2} °C. CM)
Specific heat, @ 100°F., Btu/lb., °F. or Cal/gm.°C.	0.39	0.42
Isothermal secant Bulk modulus, 77°F.	340,000 psi	328,000 psi
Foaming (ASTM D892-46T)	Essentially nonfoaming	Essentially nonfoaming
Shear stability	Comparable to MIL-H-5606A	Exceeds MIL-H-5606A
Surface Tension, 77°F. (25°C.)	30.8 dynes/cm.	28.9 dynes/cm.
Hydrolytic stability	Skydrol 500A and Skydrol 7000 are not seriously affected by low concentrations of water (less than 2%), but water contamination should be avoided.	
Refractive index, n$_D$ 25°C.	1.470 to 1.475	1.5067 to 1.507
Heat of Combustion	12,900 Btu/lb.	Not available
Thermal coefficient of expansion	0.000452/°F. 0.000813/°C.	0.000418/°F. 0.000753/°C.
Moisture % (Karl Fischer)	0.05	0.05
Dielectric Strength (ASTM D877-49) 25°C.	12 KV	36 KV
Dielectric constant 25°C., 1 KC.	8.81	8.87
Volume resistivity (ASTM D-1169) OHM-CM 25°C.	43×10^6	500×10^6

VOLATILITY

The low volatility of Skydrol 500A and Skydrol 7000 assure low evaporation rates in the hydraulic system. Both fluids are relatively unaffected by changes in pressure and temperature; Skydrol 500A and Skydrol 7000 have far lower volatility than MIL-H-5606A or similar products.

PRESSURE VISCOSITY

The change of fluid viscosity with pressure will alter the fluid flow characteristics which could slow system response. Skydrol fluids simplify this design engineering problem, for pressure has relatively little effect on these fluids.

LUBRICITY

The experience of millions of flight hours on virtually every type of aircraft hydraulic pump have proven that Skydrol 7000 generally increases the service life of most pumps. Skydrol 7000 in Douglas cabin superchargers is approved for 4000 hours of operation as compared to only 250 hours for petroleum oil. Experience with Skydrol 500A since its introduction in 1956 offers ample evidence that it too is an excellent lubricant.

CHEMICAL STABILITY

Skydrol 500A and Skydrol 7000 are heat stable and resist oxidation at temperatures beyond those encountered in actual service. In hydraulic systems, the upper temperature limit for continuous operation for Skydrol 500A and Skydrol 7000 is approximately 225°F. (107°C). Portions of the system can operate for a short time at higher temperatures without excessive deterioration of the fluids. At operational pressures of 3000 psi, Skydrol fluids have shown no tendency to thicken or form sludge over thousands of service hours.

NON-CORROSIVENESS

Skydrol 500A and Skydrol 7000 are phosphate ester-based fluids which have little corrosive effects on the metallic parts of the hydraulic system — a fact proven by years of flight experience. In general, the Skydrol fluids act as metal passivaters, which means they may be used as the preservative fluid in hydraulic components.

Table I shows the low corrosive effects of both Skydrol fluids to common metals.

Table I   Corrosion Test Results (Per MIL H-5606, 168 hrs. at 250°F. (121°C.)

| | EFFECT ON METALS (WEIGHT CHANGE MG./CM°) | |
Metal	Skydrol 500A	Skydrol 7000
Cu	−0.03	−0.92
Fe	−0.01	0.15
Al	0.00	0.02
Mg	−0.01	−0.12
Cd/Fe	−0.01	0.00
Fluid evaporation	1.0%	2.05%
Fluid separation	None	None
Color change	Fluid darkens slightly	Changes to darker green

FOAMING TENDENCY

Both Skydrol 500A and Skydrol 7000 have a very low foaming tendency. The ability of these fluids to resist air "pick-up" and reject entrained air add greater reliability of the hydraulic system.

HEAT TRANSFER

Skydrol 500A and Skydrol 7000 have good heat transfer properties. Table II lists thermal data.

Table II Thermal Data of Skydrol 500A and Skydrol 7000

PROPERTY	SKYDROL 500A	SKYDROL 7000
Specific Heat		
90 to 120°F.	—	0.45 Btu/lb./°F.
(32 to 49°C.)		
−40°F. (−40°C.)	0.34 Btu/lb./°F.	—
75°F. (24°C.)	0.38 Btu/lb./°F.	—
145°F. (63°C.)	0.41 Btu/lb./°F.	—
212°F. (100°C.)	0.44 Btu/lb./°F.	—
Thermal Conductivity		
82°F., Btu/Hr. Ft.^{2} °F. Ft.	0.0777	0.0723
28°C., Cal/Sec. Cm.^{2} °C. CM	29.9×10^{-5}	32.2×10^{-5}
178°F., Btu/Hr. Ft.^{2} °F. Ft.	0.0779	0.0716
81°C., Cal/Sec. Cm^2 °C. CM	32.2×10^{-5}	29.6×10^{-5}

FIRE RESISTANCE

Skydrol 500A and Skydrol 7000 are much less susceptible to ignition than MIL-H-5606A, as shown by tests conducted according to AMS 3150. These tests were designed to evaluate the fluids under actual operating conditions to realistically pinpoint their fire-resistance value.

In specific tests, high-pressure sprays of Skydrol fluids through the white heat of a welder's torch (often above 6000°F.) does not cause burning, but only occasional flashing. In the same test, MIL-H-5606A ignites instantly and continues burning. On a red hot manifold at 1300°F., Skydrol fluids do not burn. In other tests simulating hot manifolds, sparks, exhaust flames or electrical arcing, Skydrol fluids do not support fire. Even though they might flash at exceedingly high temperatures, Skydrol fluids could not spread a fire because burning is localized at the source of heat. Once the heat source is removed or the fluid flows away from the source, no further flashing or burning can occur because of the self-extinguishing features of the Skydrol fluids.

COMPATIBILITY OF FLUID MATERIALS

Mineral Oil. Although Skydrol fluids are miscible with mineral oil, the mixing of Skydrol with mineral oil must be avoided to maintain Skydrol fire-resistant performance. Mineral oil will seriously degrade the ability of Skydrol to resist combustion and fire, the vital justification of equipping systems with Skydrol. Similarly, mineral oil, mixed with a Skydrol fluid, will degrade seals and packings used in Skydrol-equipped hydraulic systems.

Additives used in mineral oils can also damage Skydrol systems components; for example, some viscosity index (VI) improvers used with mineral oils may not be soluble in Skydrol. In this case, the VI improver may precipitate from the hydraulic fluid mixture, leaving a gum-like residue which may interfere with proper operation of valves and filters.

Silicone and Silicate Fluids. Mixture of these fluids with a Skydrol fluid should be avoided for the same reasons mentioned for mineral oil.

Turbo Oil. Both Skydrol 500A and Skydrol 7000 are miscible with Turbo Oil 15 and 35 in concentrations up to 50 per cent; however, mixing of the fluids is not recommended, since such mixtures can degrade the fire-resistant properties of Skydrol and the packings and seals used in Skydrol equipped systems.

Miscibility of Skydrol 500A and Skydrol 7000. Skydrol 500A and Skydrol 7000 are completely miscible, and the major effect of mixing the two fluids is to increase the viscosity of Skydrol 500A at low temperatures.

ELECTRICAL PROPERTIES

Both Skydrol 500A and Skydrol 7000 possess good dielectric properties. If hydraulic system leaks develop, the Skydrol fluids do not cause short circuits, nor will they cause electrolytic corrosion in hydraulic systems. With the hundreds of miles of electrical wiring and the hundreds of circuits in today's aircraft, this lack of conductivity is another safety feature of Skydrol 500A and Skydrol 7000, which adds to their importance for hydraulic systems.

Exhibit 4 — Jack Wireman (Part 1)

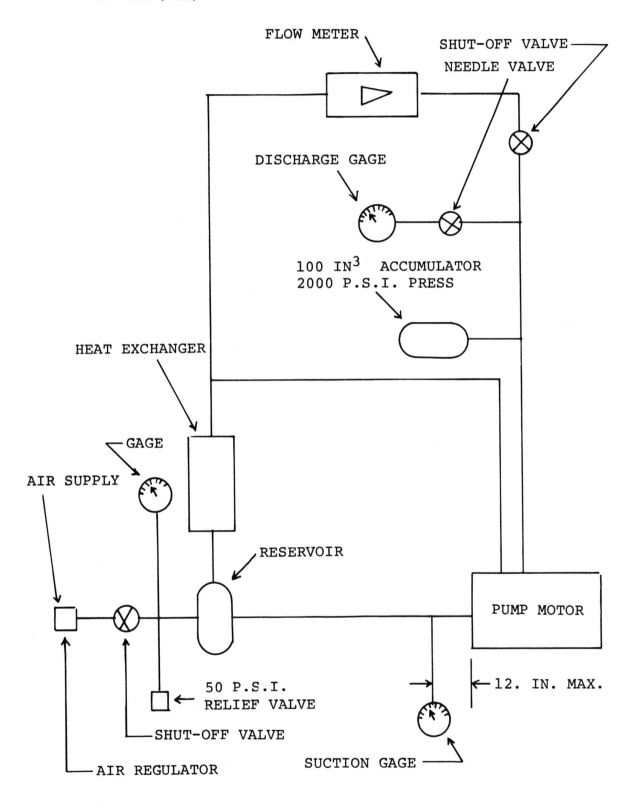

Exhibit 5 — Jack Wireman (Part 1)   Endurance-test schedule

HOURS	OUTPUT	FLUID INLET PRESSURE (MIN.)	FLUID INLET TEMP. ±5°F
225	Cycle A	5 psig	150°F
325	Cycle B	45 psig	150°F
110	1/2 gpm	45 psig	150°F
1820	1 gpm	45 psig	150°F
20	1 gpm	5 psig	Room Temp.

CYCLE A			CYCLE B		
Motor	Flow	Time	Motor	Flow	Time
On	5.7 gpm minimum*	15 min.	On	5.7 gpm minimum*	6.25 min.
On	0	30 min.	On	2 gpm	5 min.
Off	—	15 sec.	On	1/2 gpm	5 min.

* Flow which produces a motor current draw of 38.5 amperes or 5.7 gpm minimum, whichever occurs first.

Exhibit 6 — Jack Wireman (Part 1)

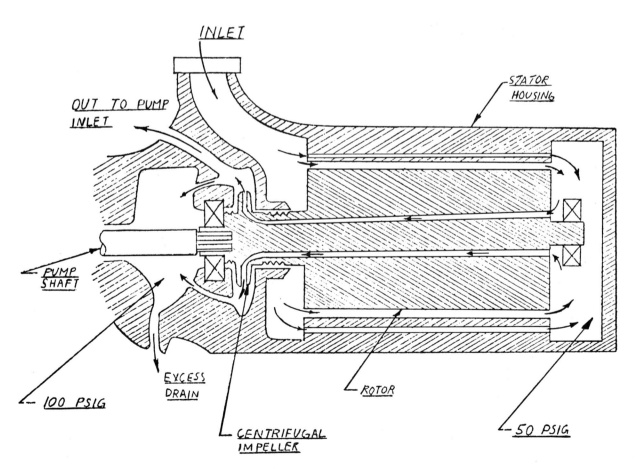

Exhibit 7 — Jack Wireman (Part 1)   Orbit Bearing catalog excerpts

For essay applications final bearing selection can be made from dimensional and performance data shown on the bearing listing pages of this catalog. Where an application runs continuously at speed, or where static load capacity is a major factor, bearing selection can be finalized only by computation of life and load factors.

Fatigue life of ball bearings that run continuously at speed under appreciable loading is considered to be the number of hours or revolutions that the bearings run before the first evidence of ball or raceway spalling develops. Fatigue is directly related to bearing load and speed. Calculation of fatigue life is of primary importance when selecting bearings for use in power-driven devices such as motors, generators, gear drives, turbines and similar equipment.

Bearing life may be limited by factors other than fatigue life, such as lubricant exhaustion, contamination, misalignment or improper fitting, and thermal constraint. In some cases, static capacity may be more important than fatigue life in selection of a bearing, as in lightly loaded components such as gyro gimbals, synchros and computer gear trains where little sustained speed is involved.

STATIC LOAD RATINGS

The static load capacity of a bearing is the peak load that can be sustained without appreciable permanent effect on smoothness of operation. Static load ratings shown on the listing pages of this catalog are based on load rating evaluation methods developed by the Anti-Friction Bearing Manufacturers Association. For most precision application, a reasonable limit value for permanent indentations is one ten-thousandth of ball diameter.

In general, less damage is done to bearings if loads are imposed while the bearings are rotating than when at rest. Even in high speed operation, however, if the load duration is so short that there is no rotational overlap of the most heavily stressed ball contact areas, separate permanent indentations may result. Bearings used in high speed applications where smooth operation is important, such as gyro rotors, should be selected so that maximum peak or shock loads are within static ratings.

In many cases, a simple statement of g loading is not sufficient for use in bearing selection. Transient and vibratory loads must also be considered, but they are frequently difficult to determine, since resonance and damping characteristics of the complete rotating assembly supports will greatly affect peak loadings imposed on the bearings. Complete vibration analysis or qualification tests under actual vibratory conditions may be needed.

BEARING SELECTION FOR STATIC CAPACITY

Pure radial loads on the bearing tentatively selected can be checked directly against the radial load rating in column C of the static load table on the appropriate listing page. Pure thrust loads can be checked against the thrust load columns T under the radial play range selected. Combined radial and thrust loads must be computed in terms of their radial and thrust load components, with bearing selection made so that each of the component loads is within its respective static radial or thrust rating. If either radial or thrust loading rating is lower than the peak load expected, a bearing with higher static rating should be chosen.

Static load ratings for DT (tandem) duplex pairs are double thrust and radial ratings of single bearings. With usual preload, the thrust rating of a **DB** (back to back) or **DF** (face to face) pair is equal to that of a single bearing but radial capacity is double that of the single bearing.

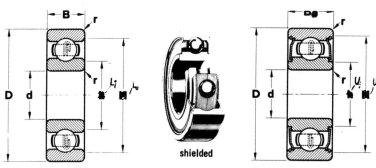

shielded

LOW TO ULTRA HIGH SPEED

DEEP GROOVE STEEL OR PHENOLIC RETAINERS

DIMENSIONS BORE, O.D. WIDTH						MAX FILLET RADIUS inches	BASIC	ORDERING	NUMBER				OTHER DIMENSIONS inches				APPROX. WIEGHT pounds	DATA REFERENCE NO.
							STEEL RETAINER			PHENOLIC RETAINER								
d		D		B				Single	Double		Single	Double						
mm	inches	mm	inches	mm	inches	r	Open	shield	shield	Open	sheild	shield	L_i	L_o	U_i	U_o		
17	.6693	35	1.3780	10	.3937	.012	103K	103S	103SS	103T	103ST	103SST	.895	1.153	.835	1.215	.09	103
17	.6693	40	1.5748	12	.4724	.025	203K	203S	203SS	203T	203ST	203SST	.952	1.292	.890	1.372	.20	203
20	.7874	42	1.6535	12	.4724	.025				104T	104ST	104SST	1.050	1.390	.989	1.458	.18	104
20	.7874	47	1.8504	14	.5512	.040				204T	204ST	204SST	1.130	1.530	1.060	1.610	.20	204
25	.9843	47	1.8504	12	.4724	.025				105T			1.247	1.587			.21	105
25	.9843	52	2.0472	15	.5905	.040	205K	205S	205SS	205T	205ST	205SST	1.320	1.720	1.250	1.800	.30	205
30	1.1811	55	2.1654	13	.5118	.040				106T	106ST	106SST	1.511	1.859	1.451	1.949	.30	106
30	1.1811	62	1.4409	16	.6299	.040				206T	206ST	206SST	1.580	2.060	1.500	2.200	.50	206
35	1.3780	62	1.4409	14	.5512	.040	107T	107S	107SS	107T	107ST	107SST	1.710	2.110	1.620	2.190	.35	107

APPLICATIONS. Motors, generators, aircraft accessories, gear drives, pumps, power tools, compressors, machine tool spindles, stable platforms, magnetic recording devices and other low to ultra high speed aplications.

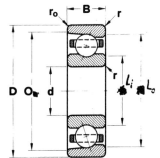

HIGH TEMPERATURE MODERATE TO ULTRA HIGH SPEED

ANGULAR CONTACT BRONZE RETAINER

Angular Contact Bearing   **Duplex Pair**

DIMENSIONS BORE, O.D. WIDTH						MAX FILLET RADIUS inches		BASIC ORDERING NUMBER	OTHER DIMENSIONS inches			APPROX. WEIGHT pounds	DATA REFERENCE NO.
d		D		B									
mm	inches	mm	inches	mm	inches	r	r_o		L_i	L_o	O		
17	.6693	40	1.5748	12	.4724	.025	.015	203HJB	.952	1.292	1.394	.18	203J
20	.7874	42	1.6535	12	.4724	.025	.015	104HJB	1.050	1.390	1.474	.17	104J
20	.7874	47	1.8504	14	.5512	.040	.020	204HJB	1.130	1.530	1.649	.30	204J
25	.9843	47	1.8504	12	.4724	.025	.015	105HJB	1.247	1.587	1.673	.20	105J
25	.9843	52	2.0472	15	.5906	.040	.020	205HJB	1.320	1.720	1.840	.35	205J
30	1.1811	55	1.1654	13	.5118	.040	.020	106HJB	1.511	1.869	1.978	.30	106J
30	1.1811	62	2.4409	16	.6299	.040	.020	206HJB	1.580	2.060	2.203	.55	206J
35	1.3780	62	1.4409	14	.5512	.040	.020	107HJB	1.710	2.110	2.229	.40	107J

APPLICATIONS. Aircraft accessories, turbines, compressors, pumps and other high speed components operating at high temperatures.

SPEED CAPABILITIES

Speed capability of bearings is almost impossible to determine exactly because of the wide variety of environmental, design and life requirements. However, the approximate speed limits given in the dynamic load rating charts on the listing pages of this catalog will serve for general guidance.

Bearings with phenolic or bronze ball retainers have highest speed capability; bearings with pressed steel retainers have lowest speed capability. Speeds for bearings with phenolic or bronze retainers are limited by type of lubrication or by outer ring centrifugal ball loading; speeds for bearings with steel retainers are limited by retainer performance.

Bearings with phenolic or bronze retainers have highest speed-life capability when lubrication is supplied during operation by continuous oil spray or mist. Prelubrication with grease gives lower speed-life capability. Lowest speed-life capability results when bearings are prelubricated with instrument oil and not supplied with additional lubricant during operation.

FATIGUE LIFE

For bearings that are properly mounted, lubricated and protected, fatigue life may be estimated as the number of hours or revolutions at a given speed and load that a group of similar bearings will operate before the first evidence of spalling or flaking of raceways or balls becomes apparent.

Design fatigue life is considered to be estimated life which will be exceeded by 90% of a group of identical bearings operating under identical load conditions assuming proper mounting, lubrication and protection against foreign substances. Average fatigue life is approximately five times this figure.

Moderate changes in the load applied to the bearing have a pronounced effect on fatigue life. Halving the bearing load increases life eight times, but doubling the load reduces life to one-eighth.

BEARING SELECTION FOR DYNAMIC CAPACITY

Equivalent Radial Load Computations. Fatigue life computations are based on two assumptions: a constant direction radial load and a condition of stationary housing (outer ring) and rotating shaft (inner ring). Therefore, combined radial and thrust loads must be converted into an equivalent radial load. Also, even where there is zero thrust load, an equivalent radial load must be calculated to take account of any radial load that tends to concentrate on a small portion of the inner ring raceway. Such an effect is developed by dynamic unbalance loading on a rotating inner ring or by a dead weight or constant direction load on a stationary inner ring. The formulas below for equiva-let radial loads take these factors into consideration.

Equivalent radial loads for individually mounted bearings are found by solving both Formulas 1 and 2 below. The larger of the two values derived is then used in subsequent formulas to find fatigue life.

Formula 1 $P = R_h + 1.2R_s$

Formula 2 $P = X(R_h + 1.2R_s) + YT$

where: P = equivalent radial loads, pounds

             R_h = radial load fixed in relation to outer ring, pounds

             R_s = radial load fixed in relation to inner ring, pounds

             X = radial load factor

             Y = thrust load factor

             T = thrust load, pounds

Identification of radial loads R_h and R_s is found in Table 7.

Table 7 Radial Load Symbols R_h and R_s

COMPONENT ROTATION	NATURE OF RADIAL LOAD	SYMBOL
Housing (outer ring) stationary, shaft (inner ring) rotating	Dead weight or constant direction load fixed in relation to housing (outer ring)	R_h
	Dynamic unbalance or rotational load fixed in relation to shaft (inner ring)	R_s
Shaft (inner ring) stationary, housing (outer ring) rotating	Dead weight or constant direction load fixed in relation to shaft (inner ring)	R_s
	Dynamic unbalance or rotational load fixed in relation to housing (outer ring)	R_h

Factor X is found by using the DATA REFERENCE NUMBER of the bearing tentatively selected to enter Table 8 to find the contact angle for the radial play range chosen. This contact angle is then entered in Chart A to determine factor X.

When appreciable thrust loads are involved, select the highest possible radial play range for the highest resulting contact angle, since both factors X and Y decrease as radial play is increased.

To find factor Y first note the appropriate value ZD^2 in Table 8 and compute value T/ZD^2. Select Chart B or Chart C as shown by Table 8 and enter with T/ZD^2 and the contact angle of the bearing to determine factor Y.

Table 8 Values for Fatigue Life Computation Spindle and Turbine Bearings

DATA REFERENCE NUMBER	CHART USED FOR FACTOR Y	INITIAL CONTACT ANGLE — DEGREES				BALL COMPLEMENT			BASIC DYNAMIC LOAD RATING C POUNDS
		Radial Play Range				Number	Diameter	Value	
		Code 3	Code 4	Code 5	Code 6	Z	D	ZD^2	
200J	C	11		13	16	9	7/32″	.431	1230
201	C			16	21	7	15/64″	.385	1180
201H	CC			13	16	9	15/64″	.495	1400
201J	C			13	16	10	15/64″	.549	1500
202	C	10		15	20	7	1/4″	.438	1340
202H	C			13	16	10	1/4″	.625	1700
202J	C			13	16	10	1/4″	.625	1700
203	C	10		15	20	8	17/64″	.565	1650
203H	C			12	15	10	17/64″	.706	1920
203J	C			12	15	11	17/64″	.776	2050
204	C	10		14	18	8	5/16″	.781	2570
204H	C			12	15	10	5/16″	.976	2740
204J	C			12	15	11	5/16″	1.07	2740
205	C	10		14	18	9	5/16″	.879	2420
205H	C			12	15	11	5/16″	1.07	2760
205J	C			12	15	13	5/16″	1.27	3090

When Y has been established, Formulas 1 and 2 may be computed and the larger resulting value taken to Formula 3 to find the design fatigue life of the bearing selected.

EXAMPLE — EQUIVALENT RADIAL LOAD COMPUTATION

Application	high speed turbine
Operating speed	80,000 rpm
Rotating member	shaft (inner ring)
Lubrication	oil spray or mist
Dead weight radial load	4.3 pounds
Dynamic unbalance at operating speed	2.3 pounds
Thrust from turbine	15.0 pounds
Thrust from preload spring	8.0 pounds
Bearing tentatively chosen	38H
DATA REFERENCE NUMBER	38H

From Table 7:

Dead weight radial load of 4.3 pounds = R_h
Dynamic unbalance of 2.3 pounds = R_s

Total thrust T = 15.0 + 8.0 = 23.0 pounds

Using formula 1: $P = R_h + 1.2R_s = 4.3 + 1.2(2.3) = 7.1$ pounds

To obtain values X and Y for Formula 2, enter Table 8 with DATA REFERENCE NUMBER 38H to find the contact angle of the bearing. The code 6 radial play range shows a contact angle of 17° as compared with 14° for code 5. Selecting code 6 for the higher contact angle and entering Chart A with 17°, factor X = 43. Entering Table 8 with reference 38H, ZD² = .220

$$\text{Value } \frac{T}{ZD^2} = \frac{23}{.220} = 105$$

Entering Table 8 with 38H, select Chart C as shown. Entering Chart C with 105 for $\frac{T}{ZD^2}$ and 17° for contact angle, factor Y = 12.

Using Formula 2:
$$\begin{aligned} P &= X(R_h + 1.2R_s) + YT \\ &= 43(4.3 + 2.8) + 1.2(23) \\ &= 3.1 + 28 = 31.1 \text{ pounds} \end{aligned}$$

FATIGUE LIFE COMPUTATIONS — HOURS

Since fatigue life is usually computed in terms of hours at a given speed, the dynamic load ratings C_s are given in the dynamic load charts on the listing pages of this catalog for speeds of 100 rpm and higher. The design life basis for C_s values is 500 hours for 90% survival, as developed from load rating evaluation methods standardized by the Anti-Friction Bearing Manufacturers Association.

$$L_{10} = 500 \left(\frac{C_s}{P}\right)^3$$

where:
L_{10} = life in hours
C_s = dynamic radial load rating
P = equivalent radial load, pounds

EXAMPLE — FULL LIFE COMPUTATION

$$L_{10} = 500 \left(\frac{C_s}{P}\right)^3 = 500 \left(\frac{43}{31.1}\right)^3$$

$$= 500(1.38)^3 = 1314 \text{ hours}$$

CHART A – RADIAL LOAD FACTOR X

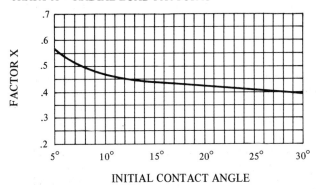

INITIAL CONTACT ANGLE

CHART B – THRUST LOAD FACTOR Y OR Y_d (SEE TABLE 8)

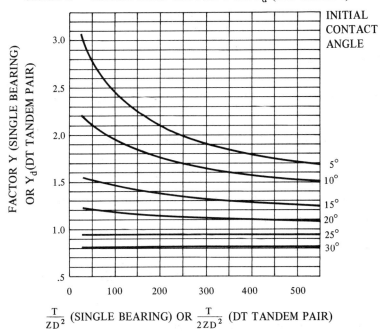

$\dfrac{T}{ZD^2}$ (SINGLE BEARING) OR $\dfrac{T}{2ZD^2}$ (DT TANDEM PAIR)

CHART C – THRUST LOAD FACTOR Y OR Y_d (SEE TABLE 8)

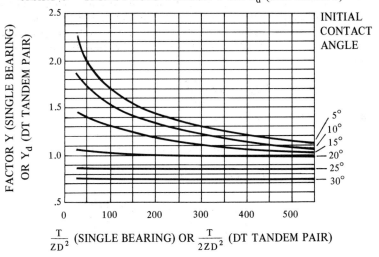

$\dfrac{T}{ZD^2}$ (SINGLE BEARING) OR $\dfrac{T}{2ZD^2}$ (DT TANDEM PAIR)

Dynamic Load Ratings. Radial load ratings below are for 500-hour design life, or 2500-hour average life, with proper mounting, lubrication and protection of bearings. Use value C_s for fatigue life computation procedure.

○ Normal speed limit for open or shielded bearings with steel retainers.

□ Normal speed limit for open or shielded bearings with phenolic retainers, grease lubrication.

△ Normal speed limit for open bearings with phenolic retainers, oil mist or spray lubrication.

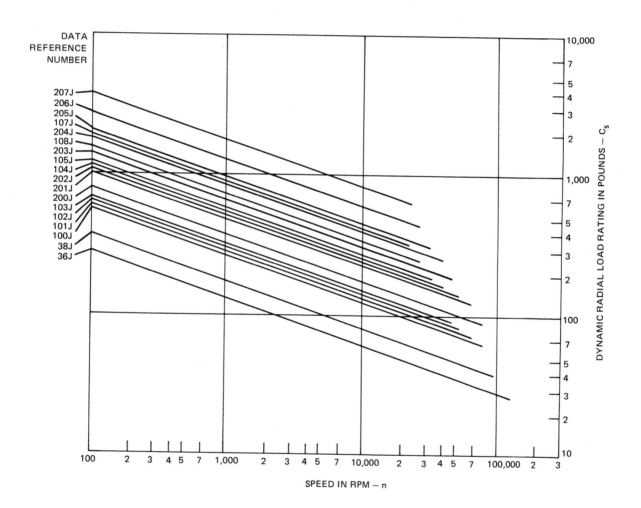

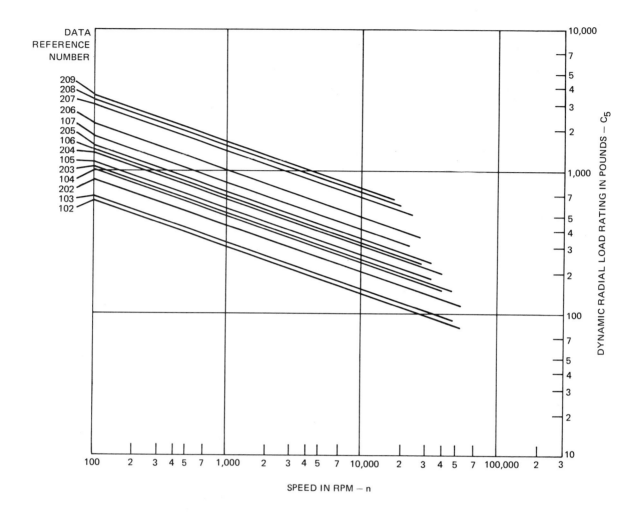

STATIC LOAD RATINGS — POUNDS

DATA REFERENCE* NUMBER	THRUST LOAD — T						RADIAL LOAD C_s
	RADIAL PLAY RANGE						
	Code 3		Code 5		Code 8		
	.0002″	.0004″	.0005″	.0008″	.0008″	.0011″	
102	2030		2120		1800		650
103	2300		2350		1700		730
	.0002″	.0005″	.0005″	.0009″	.0009″	.0014″	
202	2100		2840		2380		770
203	2720				3000		1000
104	2180		2210		2340		900
204	3900		4000		4160		1400
105	2000		2300		2400		950

Static Load Ratings. Static ratings are shown . . . for radial play ranges normally supplied. These values indicate peak loads that can be sustained by bearings without permanent effect on smoothness of operation.

LUBRICANTS AND LUBRICATION

Proper and adequate lubrication of precision ball bearings is essential if maximum performance is to be obtained. Of equal importance is the exclusion of foreign matter by suitable design of enclosure and, where possible, the use of shielded bearings, even for the most critical low torque applications.

Lubricants normally supplied are shown on the listing pages. General guides for oil or grease lubrication are presented in Tables 9 and 10. These tables list in summary form the oils and greases most frequently specified, along with their temperature and speed ranges. Other lubricants, some of which are being used successfully under more exacting conditions, can be supplied on request.

Oil. With oil lubrication, performance life is mainly dependent on temperature and speed since higher temperatures increase the rate of evaporation and oxidation, and higher speeds tend to dissipate the oil. Extreme caution should be observed in specifying minimal amounts of oil because of shortened life resulting from lubricant starvation.

Grease. Moderate speed and heavier load applications operate successfully with grease, which is more readily retained than oil within the bearing or housing enclosure. Grease tends to remain in close proximity to the critical load carrying areas and provides lubricating action by bleed-out of oil in the grease. Performance life in such cases usually is limited by oxidation and thermal stability of greases and evaporation of oil content. Air passage through bearings has an adverse effect on greases and bearing life and therefore should be minimized.

Bearings are factory lubricated with grease by filling 20 to 35% of available internal space. Standard quantities meet the needs of most applications. Reduced quantities for unusual requirements can be supplied on special order.

Vacuum Impregnated Retainers. Excellent results are attained in certain applications by vacuum impregnation of phenolic ball retainers with oil and subsequent addition of grease. All oils and greases are not compatible, but in general the oil component of the grease should be the same type as the oil used for impregnation.

Oil Spray or Mist. Advantages of pressurized air-oil mist equipment for machine tool spindle applications include once-through passage of clean oil, superior cooling of critical bearing surfaces and exclusion of foreign substances. The oil mist should be directed from the interior of the spindle housing as exhausted outward through the housing seals.

Pressurized air-oil mist systems require a source of clean compressed air, thus are usually too cumbersome for airborne or space vehicles. Spray oil or wick fed systems may be used for many airborne applications but specially designed prelubricated or self-lubricating bearings may be required for space vehicles.

SHAFT AND HOUSING FITTING PRACTICE

Improper fitting of precision bearings is almost certain to result in poor performance. If the bearing seat is out of round or if the fit is too tight, dimensional accuracy of the raceway may be seriously disturbed, or the carefully controlled internal clearance in the bearing may be destroyed. Too loose a fit, on the other hand, may prevent proper functioning because of lack of control of the rotating component.

Interference fits should be approached with caution, particularly with thin section bearings, as they can cause loss of internal clearance or force the raceway out of round with resulting noisy or poor operation. Close clearance fits are usually recommended to avoid distortion or damage to bearings, but they result in looseness that may be objectionable for certain applications. This effect can be eliminated by suitable face clamping or locking means.

While an interference fit will prevent fretting corrosion and wear of the seating surface through relative movement, only one ring of a bearing should normally be interference fitted. The other ring should be free to move axially on the housing or shaft seat to relieve thermal expansion or contraction forces or to permit proper axial adjustment or spring preloading.

Problems are sometimes encountered with outer rings turning in stationary housings. This is usually due to rotating radial loads created by dynamic unbalance. Better balancing will usually correct such a condition, providing a sounder solution than interference fitting of the outer ring in the housing.

High speed turbine applications also may require more clearance in the housing, particularly where some heat is transmitted from the shaft through the bearing in addition to heat generated within the bearing due to the high speed.

Although soft non-ferrous housings may be used for lightly loaded applications at normal temperatures, steel housing liners should be used for all applications where there is a wide range of operating temperatures. A steel liner will resist contraction of the housing on the outer ring of the bearing at low temperatures and minimize excessive loosening of the fit at high temperatures.

Examples of Shaft and Housing Fits. Examples of typical shaft and housing fits, based on working tolerance ranges on both bearings and mounting parts, are shown in Table 11.

Table 9 Guide to Oil Lubrication

APPLICATIONS	TEMPERATURE RANGE	MILITARY OR OTHER SPEC.	COMMENTS OR LIMITATIONS*
Gyro gimbals, synchros, computer gear trains, other low torque, low speed applications	−70° to +175°F	MIL-L-6085A (Anderson-Winsor L245X)	Light general purpose diester instrument oil. Good rust prevention properties. Low starting and running torque dn — 20,000 max. for pre-lubricated bearings.
	−100° to +350°F	G.E.-Versilube F44	A light silicone oil with oxidation inhibitor. Very broad temperature range. dn—20,000 max. for prelubricated bearings.
Gyro rotors of inertial gyro systems	+140° to +200°F	Esso-Teresso V-78	A highly refined paraffin mineral oil, vacuum impregnated into phenolic or other oil absorbent retainers in spin axis bearings for minimum oil mass shift. Used alone or preferably with G-6 grease. dn — 250,000 max.
High temperature, high speed aircraft turbines, drives, pumps, other accessories with heavily loaded gears	−67° to +400°F	MIL-L-7808C (Esso-Turbo oil No. 15)	A diester gas turbine oil normally applied by oil spray or mist system. For low to high speed, high temperature service.
High temperature, high speed aircraft turbines, pumps, compressors	−67° to +400°F	MIL-L-6085A (Anderson-Winsor L245X)	An excellent low to high speed, high temperature oil when applied as spray or mist. Better than 0–14 unless gears must also be lubricated. Less coking than 0–14.

* dn = bearing bore in mm (d) × speed in rpm (n) — for inner ring rotation only; if outer ring rotates, use one-half the value shown.

Table 10   Guide to Grease Lubrication

APPLICATIONS	TEMPERATURE RANGE	MILITARY OR OTHER SPEC.	COMMENTS OR LIMITATIONS*
Gyro rotors, small motors, generators, computers and similar applications	−65° to +200°F	MIL-G-3278A (Esso-Beacon 325)	An excellent general purpose smooth running instrument bearing lubricant. Low volatility, long life. Lithium soap, diester oil. To 400,000 dn. Useful to 250°F for short periods.
Gyro rotors, motors, generators, machine tool spindles, power tools	−25° to +250°F	Esso-Andok C	A stiff grease which channels readily. Mineral oil, sodium soap. Good mechanical stability, low mass shift. Excellent for outer ring rotation. Smooth running with good load carrying ability. Excellent high speed grease. To 600,000 dn.
Small motors, generators, computers, gear trains	−100° to +350°F	Dow Corning-DC 33 light	A silicone grease with lithium soap. Useful under light loads (1/3 catalog rating at speed) over a wide temperature range and at low speeds (dn — 200,000 max.).
Gyro rotors, motors, generators	−65° to +350°F	Texas-Unitemp 500	A sodium organic thickened semi-fluid grease using synthetic oils. Broad temperature range and good speedability (dn — 500,000 max.). Feels gritty in bearing before run-in. Torque high at −65°F.
Motors, generators	−40° to +450°F	Shell-ETR-B	A silicone grease with organic dye thickener. Broad temperature range but with low speedability (dn — 200,000 max.) and light load carrying ability (1/3 catalog rating at speed). Feels gritty before run-in.
Gyro rotors, motors, generators	−65° to +350°F	MIL-G-25760 (American Oil-Supermil 06752)	A synthetic grease with aryl-urea thickener. Good speedability (dn — 400,000 max.) and very broad temperature range. Smooth running. Longer life than G-2.
Preservative coating for storage	+40° to +150°F	MIL-C-11796A-Class 3 (Anderson)	A preservative slush. If bearings are subsequently lubricated with oil mist or spray, preservative need not be removed.

* dn = bearing bore in mm (d) × speed in rpm (n) — for inner ring rotation only; if outer ring rotates, use one-half the value shown.

Table 11 Shaft and Housing Fits

			SPINDLE AND TURBINE BEARINGS	
		Dominant Requirements	*Fit Extremes — Inches**	*Typical Applications*
Shaft fits	Inner ring clamped	Very low runout, high radial rigidity	+.00005 (tight) −.00025 (loose)	Grinding spindles, magnetic drums, platform gimbals
		Low to high speeds, low to moderate radial loads	+.00005 (tight) −.00025 (loose)	Turbines, compressors, motors, generators
		Heavy radial load	Inner ring rotates	+.00005 (tight) +.00035 (tight)	Gear or belt loaded applications
			Outer ring rotates	.0000 −.0003 (loose)	Idler or planet gears
	Inner ring not clamped	Very low runout, high radial rigidity	+.00005 (tight) +.00035 (tight)	Magnetic drums, grinding spindles, platform gimbals
		Moderate to high speeds, light to moderate radial loads	+.00005 (tight) +.00035 (tight)	Turbines, compressors, motors, generators, gear shafts
		Heavy radial load, low to moderate speeds	Inner ring rotates	+.00005 (tight) +.00035 (tight)	Gear or belt loaded applications
			Outer ring rotates	.0000 −.0003 (loose)	Idler or planet gears
		Inner ring must float to allow for expansion, low speed only	−.00005 (loose) −.00035 (loose)	Platform gimbals
Housing fits	Normal accuracy, low to high speeds, moderate temperature		.0000 −.0004 (loose)	Spindles, magnetic drums, motors, generators
	Very low runout, high radial rigidity		+.0001 (tight)	Platform gimbals, resolvers, inductosyns
	Outer ring need not move readily to allow for expansion		−.0003 (loose)	
	High temperature, moderate to high speed		−.0001 (loose)	Turbines, compressors, starters
	Outer ring can move readily to allow for expansion		−.0005 (loose)	
	Heavy radial load Outer ring rotates		+.0002 (tight) −.0002 (loose)	Idler or planet gears, cam rollers

* tight fits are positive (+) and loose fits negative (−).

Three CNCEPTUAL DESIGN

Having defined a problem, the engineer's task is to find a good way to solve it, and this calls for identification of possible alternative designs. As with the definition of the problem, the creation of a solution concept usually allows many possible alternatives, and there may be a trap in the human tendency to grasp the first solution idea that comes to mind and cling to it. If the problem is a minor one, such as which kind of soft drink to select from a dispenser, then the marriage-at-first-sight approach may be clearly the most economical use of intellectual energy in arriving at a satisfactory solution. But if the problem is an important one, and especially if it is new to the engineer, then there is real danger in embracing the first idea before considering other possibilities. There may be a better idea that will appear in a competitor's product and outclass the first-love solution. A sufficiently thorough search should be made so that whatever ideas others suggest after the engineer has selected his solution, he can say, "I considered that idea, but the one I chose turned out to be better."

Better ideas, even apparent breakthroughs, are often not large steps, but rather represent small improvements or a synthesis of already existing elements. Major innovations such as the Wankel engine and the transistor are rare, and even for these there can be traced many close links with prior discoveries that made them possible. Because of this tendency to build always from the established base, it is not unusual to find very nearly simultaneous development of new devices by engineers operating independently of each other. Others were developing workable aircraft at the same time as the Wright brothers. Langley had a workable design prior to the Wrights, although its successful flight followed theirs. Technological competition in World War II found the British and the Germans independently creating radar and jet engines at very nearly the same time.*

* For a study of how new engineering developments follow from prior art, see P. J. Booker, "Principles and Precedents in Engineering Design" (London: Institution of Engineering Designers, 1962).

Yet it must be said that the prior discoveries that make breakthroughs possible are usually available to many, and the one out of many who is first to produce the new synthesis deserves credit for doing so. In the competition for the contract to produce a new light observation helicopter for the U.S. Army during the 1960s, for example, three well-known companies entered. The stakes were high because the contract would represent the largest number of aircraft purchased under a single order since World War II. All the companies had designed and produced helicopters for years, so there was no imbalance of preparedness or resources. Moreover, the techniques of helicopter design were fairly well established, so it could have been expected that the competition would be close and the results from the three companies nearly similar.

In fact, however, one company produced a design that weighed about one-third less, could carry about one-third more, and could fly substantially higher, faster, and farther at less cost. Moreover, its price tag was significantly lower than the other two. Throughout the craft were innovations in design that gave it in aggregate a clear margin of superiority. For example, a novel arrangement for the main rotor hub used light, flexible members instead of the more cumbersome and heavier joints, which had been traditional. A single cooling fan adapted from the Porsche automobile was used to perform functions that took two separate fans and associated couplings in the other craft. Throughout the ship it was apparent that the design team had worked with extraordinary resourcefulness to outperform their competitors.

How does outstanding creativity come about? Is it produced by more brains? Or by greater effort? Or by more knowledge? Or by different attitudes? Or is it simply a matter of luck? Nobody has provided a single formula for the answer. It appears that any or all of these ingredients may play a part in any given instance, and consequently it is advisable to work on all of them. There has been considerable psychological research on the subject of creativity in the sense of ability to generate ideas, and the following conclusions have been found, supported by experiments:

1 There is no correlation between IQ and creativity in the normal range. IQ generally correlates with school ability, but it is possible to be low in this area but very creative; or high in this area but very uncreative.
2 Deliberate practice in idea generation (with such exercises as those appearing at the end of this discussion) can raise creativity-test performance.
3 Effects of practice persist with time.
4 Suspending judgment, deliberately withholding criticism (no matter how weird the idea seems), raises the quantity of ideas generated.
5 Quantity of ideas tends to beget quality. There is a value in delaying judgment while you try to generate as many ideas as possible before making a final decision.
6 Seeking cleverness in ideas tends to beget fewer but more clever ideas. So there seems to be some degree of tradeoff between quantity and quality. Where the optimum is for any given individual is unpredictable.
7 Better ideas tend to come later in idea-generating sessions. For most people it pays to work hard at generating ideas. This work may often pay off in ideas that surface only days or weeks after the subject has been set aside (in Mark Twain's words, "to let the tank fill up").
8 Many heads produce more ideas than one. It pays to seek ideas from others both individually and in brainstorming groups.

It can be noted from this list that there are some conflicting forces present. For example, quantity tends to produce quality in ideas, and the way to get quantity is to suspend judgment and generate ideas as freely as possible (produce "cerebral popcorn," as some have called it). But by imposing some judgment it appears that still higher-quality ideas can be produced, even though quantity will be reduced. So the inventor must form a judgment as to how much he will relax his judgment in seeking ideas. The fact that engineering advances appear generally to occur in relatively small steps from prior experience also suggests that allowing extreme wildness in an idea search is not very likely to pay well.

There is a commonly expressed view that it is easy to come up with ideas but hard to find good ones. To pursue all the new ideas people can come up with would be prohibitively expensive, so discretion must be applied to support only those that are most worthwhile. The question as to whose ideas will be most worthwhile largely comes back to a matter of competition, as in the example of the helicopter contract mentioned above. To win, one ground rule consistent with all the experimental findings noted above

as well as by industrial experience is that it pays to work at being creative and to be persistent about it.

Why there is need for effort and persistence to produce creative output becomes more apparent when the list of possible psychological blocks to new ideas is considered. One such list* includes the following:

1 Perceptual blocks
   a Difficulty seeing problem from different perspectives
   b Delimiting problem too closely or too conventionally
   c Seeing only what you expect or are expected to see
   d Saturation; perceptual numbness
   e Failure to use all sensory inputs
2 Emotional blocks
   a Lack of challenge or interest in problem
   b Excess zeal, tunnel vision, insufficient patience
   c Fear of failure and of risk
   d Intolerance of ambiguity, obsession with security, order
   e Preference for judging rather than seeking ideas
   f Inability to relax, incubate after strong effort on problem
3 Cultural blocks
   a Disdain for fantasy, reflection, idea playfulness, humor
   b Reason, logic, numbers superior to feeling, intuition, pleasure
   c Tradition preferable to change
   d Science and money are the way to solve all problems
   e Taboos are taboo
4 Imagination blocks
   a Fear of unconscious
   b Inhibition about some areas of imagination
   c Compulsions such as worry, order, activity
   d Confusion of reality and fantasy
5 Environmental blocks
   a Distrust among colleagues
   b Autocratic boss who doesn't recognize and reward others' ideas
   c Insecurity
   d Distractions
   e Lack of implementation support
   f Repressive group norms, such as critical or decisive responses
6 Intellectual blocks
   a Lack of information or incorrect information, knowledge
   b Ineffective application of formal idea-generation techniques
   c Misformulation of problem

* After J. L. Adams, "Individual and Small Group Creativity," *Engineering Education* 63 (November, 1972).

    d Intimidating or distracting weakness of formal skills such as math

7 Expressive blocks

    a Lack of medium facility, verbal, pictorial, physical

Working from a list such as this, a creatively inclined reader should be able to generate a companion list of antidotes for coping with the blocks. Many others have studied the creative process and reached some degree of agreement that in passing the hurdles to success there are several fairly distinct and identifiable phases. The first is preparation, in which the problem-solver equips himself with the skills and knowledge needed for tackling the problem, possibly as much or more by accident as by plan. Then comes a searching stage, in which possible solutions are sought out and considered. This leads to a third stage, which is frustration. Nothing seems to work. Then the problem-solver may give up for a while and turn to other things. At some unpredictable moment there may then come the sudden inspiration of the needed idea, somehow having been processed by the subconscious during incubation. Many inventors have recounted how inspiration struck at unlikely moments when their minds were elsewhere, such as when shaving, driving a car, or even sleeping. They also tend to agree that the preparative stages and effort leading to frustration are necessary to produce the inspiration.

It appears that five controllable factors can be influential in the creative process:

1 Attitudes that rate creative production as important and that affirm the problem-solver's capacity to be creative can be cultivated

2 Knowledge appropriate to the problems to be solved can be acquired

3 Circumstances such as job atmosphere and colleague relationships can be selected and developed to improve odds of favorable innovative outcomes

4 Methods for generating ideas — creativity tricks — can be applied at will

5 The vital ingredient of persistent effort can be deliberately brought to bear

The presence of these five ingredients will tend to foster the appearance of another: luck.

To help in idea generation, those who have studied and practiced creative proficiency have, of course, generated many suggestions, including the following:

1 Try different ways of defining the problem.

2 Work at being prolific. Schedule idea-generating practice sessions.

3 Seek new ideas from others. Use opportunities to join or organize brainstorming groups.

4 Don't be discouraged by others' negative views. Devil's advocates are easy to find. Experts in the past have predicted impossibility for such things as gas turbines, atomic power, FM radios, and airplanes.

5 Avoid stating negative views until you have generated possible solutions for the obstacles, then offer the solutions rather than the negative views. (This alone will make you much rarer than the devil's advocates, and much more valuable.)

6 Set quotas and deadlines for generating ideas, but be open to still better ideas that emerge later after more incubation.

7 Resist the temptation to accept early ideas. Keep generating more before passing judgment, then go back and evaluate them later.

8 Utilize tricks for multiplying ideas, such as elaborating on those ideas that emerge, rearranging, reversing, expanding, shrinking, combining, and altering them. Try fantasies, listing attributes of needed solutions, and combining ways to satisfy the attributes, and so on. Cast yourself in the role of the physical thing you are trying to design or improve. How does it feel? What would make it better?

9 Reach for the frontier. If you don't get as far as to ideas that make you laugh, you are being too conservative.

10 Discuss with others. Explain the problem, especially the toughest parts, to other people who are sympathetic but new to the problem. Consider it your success rather than your failure when they reveal insights into solutions you overlooked.

Although important, creativity is clearly not enough to produce successful engineering. There must also be effective analysis, decision making to choose among the creative ideas, and finally action to put the chosen ideas into effect. The value society places on follow-through can be seen in the fact that promoters, organizers, and salesmen who carry out their functions are often well rewarded, in many cases better than those who first thought of the ideas. There are many who have been creative but failed to sell or otherwise have their ideas implemented, only to see others subsequently capitalize on them. An engineer is well-advised, therefore, to arm himself with tools of presentation and persuasion, including skills in artwork, writing, speaking, salesmanship, and working with others.

Exercises

1 What other processes besides idea generation must your mind go through in designing? Discuss how time should be allocated among the different processes.

2 Leonardo da Vinci sketched flying machines.

Why did it take so long for them to be developed? What technology was an absolute prerequisite?

3 With a group of three or more other students, generate as many ideas as you can on one of the following topics in 20 minutes; assign someone to write them down.
    a Childproof pill containers.
    b What the world would need if everyone were to be without sight for one year.
    c Schemes for prevention and cleanup of oil spills.
    d The ideal house.
    e Ways of getting more ideas.
    f Individualized furniture.

4 Select some object that was designed to serve some practical function. Imagine yourself as being that object and describe what it feels like to perform the intended function. List modifications that would enable you to perform the function better.

5 In 2 minutes list as many as you can of the following:
    a Uses for a pencil.
    b What the consequences would be if it were always daylight.
    c Ways the grading system could be changed.
    d Devices that could be powered from a household water faucet.
    e Gadgets that could be added to improve automobiles.
    f Names for an improved pillow.

6 Set a target of something like six new product ideas per week and list these in a notebook. Try different schemes for generating ideas and note which ideas came from which scheme. At the end of the term, describe your views of the effectiveness of the different idea-generation schemes.

7 Visit a local firm and ask (1) to what extent they encourage employees to submit new ideas and how they encourage them; (2) what examples of such ideas they can point to and which employees they came from; and (3) what forms, if any, they require employees to sign regarding disclosure and ownership of new ideas.

8 Discuss the saying "necessity is the mother of invention."

9 Develop a preliminary design for one of the following, keeping a record of your mental design process:
    a A toy for blind children.
    b A remotely operated soccer ball.
    c An electric nutcracker.
    d A wheel balancer to cost $1.
    e An automobile lubrication-time notifier.

10 Make an evaluation matrix for colleges you might attend. List possible colleges in a vertical column down the left side of the sheet of paper, and characteristics you care about in a college in a horizontal column along the bottom at 90° to the first list. Fill in the resulting matrix with properties of the colleges. What do the results show?

11 Make an evaluation matrix as above for some existing product such as a specific automobile. List important design parameters along the left side of the page and alternative designs along the bottom. Fill in the matrix with properties of the designs and evaluate them. How do decisions depend on the judged importance of various criteria?

12 Make an evaluation matrix as in the above problems but apply it to design concepts generated in one of the earlier exercises. Discuss how such a matrix can be used to help sell a new idea.

J.L. ADAMS CORPORATION

Areas for Improvement of Functional Artificial Limbs

The company doesn't employ a professional engineer. But it knows its artificial limbs have a number of deficiencies in design. It is also aware of exotic new approaches that might be taken to other prosthetic problems as well. What could an engineer do for them? How?

In early 1965 T. R. Parke, vice president of the J. L. Adams Corporation, was shown a memorandum from the Committee on Prosthetics Research and Development of the National Academy of Sciences. This memorandum, part of which appears in Exhibit 1 — J. L. Adams Corporation, was a compilation of suggestions for design needs in the fields of artificial limbs, leg braces, and control devices for handicapped people.

"I'm not sure how much we could do to help with any of these," Mr. Parke observed. "Although our company makes the key components in over 90 percent of all artificial arms worn in the United States, we have not felt we could afford an engineer. If somebody could do further creative thinking and tell us what it would take to go ahead with some of these product possibilities, maybe we could reconsider, but the company doesn't have money for any $20,000 research and development projects unless the payoff is demonstrably pretty spectacular.

"Government contracts for work on the more exotic problems in prosthetics usually go to organizations whose main line of activity is something else, such as education, general research, or even defense production. We would like to get ideas for some of our future products from these government-sponsored prosthetic investigations, but unfortunately many of the engineers on these projects don't really understand the needs of amputees. We concentrate on these needs because our whole business is manufacturing prostheses for these people to wear.

"Although our products have been developed, tested, and proved in wide usage over more than a decade, we feel that our technical efforts should still be concentrated on improving them. The sort of question we would like to have answered, for instance, would be: How can the design of our mechanical elbows be improved? We don't want to completely ignore new-product ideas, but they will have to be pretty convincing."

The Company and Its Products

The Adams Corporation was formed during World War II to develop and produce an artificial arm with an improved wrist. The wrist idea had been conceived by Dr. Adams, a wealthy surgeon who had lost a hand as the result of an infection that developed after a minor operation performed by another surgeon. Dr. Adams' investment advisor suggested formation of a company to produce the new device, and the two men established it jointly. Further innovations were introduced, such as use of Fiberglas in forming the fitted sockets, which had formerly been carved out of solid wood. This was the conception of the investment advisor, who had also been trained as an engineer. He later bought out the doctor's interest, and then sold the company to another firm and returned to the investment counseling field.

War casualties produced a substantially growing need for their products, and the product improvements were so successful that Adams soon became the leading supplier of artificial arms. It also developed new designs for artificial legs, including mechanical knees. Following the war, sales diminished, increased during the Korean conflict, dropped again, then increased once more during the Vietnam war. As of 1965 the company employed 25 people. Its draftsman also served as model maker and in cooperation with shop and sales personnel worked on design changes. Mr. Parke doubted there was enough work to justify hiring a full-time engineer. Present products were selling well, 25 percent of sales being to 30 countries outside the United States. When the Vietnam war ended, Mr. Parke expected sales would again decline unless new products were introduced to broaden the company's product line.

The company usually did not deal directly with end users of its products. Most arms were sold through independently operated retail shops,* which custom-tailored the prostheses to individual amputees. Typically, a customer came to the shop on the referral of his doctor. He was then interviewed and measured to determine the optimum shape and strength prosthesis for his particular range of applications. Then an arm was built up from components supplied by Adams, such as elbows, cables, and wrist connections, and from plastic parts made at the fitting shop.

Because the company did not usually deal directly with customers, Adams found that product defects often did not come to its attention when they occurred. For example, the company had recently received a letter from a customer informing them that his wrist joint would often jam. The company draftsman made some tests at the plant and found that the wrist units in stock also jammed. He then called some of the fitting shops and discovered that they had been modifying the elbows with the addition of a rubber washer and had not bothered to inform Adams of the trouble.

Almost all of the 1,200 different items produced at Adams were handmade, because the company management felt it could not justify the cost of mass production tooling. "Most of the parts we make are low-production items," said Mr. Parke, "and therefore the only way we could make mass production economical would be to produce the same design for a number of years. We are reluctant to freeze our designs for the time required to pay for the tooling."

The major production facilities at the plant are listed below:

four turret lathes (max collet 2½ in. diameter)
four horizontal lathes (9- to 16- in. swing)
two vertical mills (18-in. bed-travel)
two horizontal mills (18-in. bed-travel)
one tool-room-type jig bore
plastic fabrication shop for resin and cloth laminates
one 22-ton punch press
one small spot welder
one small arc welder
tumbling equipment
one 8-drill, drill-press bank
one 5-spindle drill press

* These shops are known in the prosthetic industry as "fitting facilities."

Design at Adams

The company occasionally hired an outside engineering firm to do design work, but most parts were designed by the management and other employees. "We have a very ingenious machinist at Adams," said Mr. Parke. "Before we hired a designer, I would go to him and describe what I needed. He would 'hog' out a part he thought would do the job, and we would test it out for a while. Then I would go back and tell him what was wrong with the part. We would go through a series of these interactions until we were satisfied with our design.

"We are just beginning to make mechanical drawings of our mechanisms," said Mr. Parke. "Our machinists have usually taken their dimensions from the parts that they are duplicating. This has been a satisfactory arrangement in most cases, but we did have a problem once. We found that some of our mechanical elbows were jamming. All of the parts inside looked as if they should have worked, so it took us quite a while to track down the source of trouble. We finally found that one of our machinists had incorrectly cut a sector gear by a few thousandths of an inch. We had to throw away all of the defective gears."

Mr. Turner, who had repaired elbow units at the company for 15 yr, told of another way products were designed at Adams. "We are a small company and everyone knows everyone else. So it's easy for us to let the management know about anything we think should be improved. If I find that some part keeps breaking on the arms I repair, I tell the foreman or the president or vice president about it and suggest some ways the problem might be corrected." For each suggestion used, Mr. Parke said that Adams would reward an employee with as much as the resulting estimated saving for 1 yr.

"We designed a small beryllium-copper spring a few years ago, and the first lot we produced worked fine," recalled Mr. Turner. "However, the second lot of springs was brittle and broke after a few cycles of bending. We thought improper heat treatment might have been the problem, so we had two different heat-treating companies make some of the springs. These springs also failed after a few cycles. We never did find out what was wrong. Finally, we designed another spring with a new shape, and made these from steel instead of beryllium-copper. We've had no trouble with the new spring."

Mr. Parke believed an important asset to his company was that he had worked in the field of prosthetics prior to coming to Adams, and that he wore two artificial arms.

Upper-Extremity Prosthetic Devices

A distinguishing feature of prostheses made from Adams' components, according to the company, was that these units permitted an amputee to perform many manipulative tasks that a real arm can perform. Mr. Parke stated that it was impossible to list the deficiencies of a prosthesis as compared to a real arm since these deficiencies vary for different amputees. Adeptness in using a prosthesis is affected by such factors as the amputee's age, training, strength, and mental attitude. Exhibit 2 — J. L. Adams Corporation shows some of the tasks that a prosthesis wearer can perform. Mr. Parke added that there was one exception to his general statement. None of the existing upper-extremity prostheses provided much tactile feedback. This was not considered a great problem by his company. However, it could cause difficulties when an amputee had to hunt for a light switch in a dark room or attempt to locate change in his pants pocket.

There are many possible configurations for upper-extremity prostheses. The hand * on most of these units was a pair of opposing metal tongs called a split hook. "There are some terminal devices on the market that resemble real hands," said Mr. Parke. "However, these devices do not perform nearly as well as split hooks. Furthermore, they can cause awkward situations when someone shakes hands with a person who wears one. These hands look very realistic; however, they are cold to touch, and this can come as a shock. The terminal devices that resemble hands are a lot more expensive, on the average, than split hooks; the most popular 'cosmetic hand' sells for about $135, while the average split hook sells for about $65." One tong of the split hook is connected to a steel cable attached by straps to the amputee's shoulder. The amputee can pull on the cable by moving his shoulder. The cable in turn pulls on one of the tongs, causing it to pivot away from the other one. When the shoulder is relaxed, several rubber bands, connected between the two tongs, pull them together. Each rubber band provides 1 lb of "pinch," and five or six are typically used.

A threaded stud, protruding from one end of the split hook, fits into a socket on the wrist. Some of these wrist sockets are shown in Exhibit 3 — J. L. Adams Corporation. The socket allows the amputee to rotate his split hook about the connecting stud by turning the hook with his other hand. With some, he can also lock his hook in a fixed position, if he desires. Some wrists contain quick-change devices that allow the amputee to interchange terminal devices conveniently.

The wrist unit is laminated to a piece of plastic material corresponding to the forearm (see Exhibit 4 — J. L. Adams Corporation). The type of forearm and remaining parts in a prosthesis depend upon the length of the amputee's stump relative to the remaining joints in his arm. If the amputation

* An artificial hand was known in the prosthetics industry as a "terminal device."

was below the elbow, the plastic forearm is hinged to a plastic cuff strapped to the amputee's upper arm. If the amputee's forearm stump is long enough, it is fitted into a socket in the plastic forearm. The stump and plastic forearm then move together as a unit. If an arm were amputated at or above the elbow, other designs would be used.

A mechanism in the elbow is actuated by a cable to allow locking in several alternative positions. One pull on this cable, which is anchored to the shoulder harness, and the hinge releases. Another pull locks it again.

Plastic parts such as the forearm and upper arm sockets are usually made in plaster molds, which were made from casts of the amputee. The plastic typically consists of five laminations of woven tubular nylon cloth saturated with polyester resin. This makes a hollow shell with wall thickness ranging from ¼ to ⅜ in. Glass fiber is added where extra strength is needed, but is avoided wherever possible because of its tendency to irritate the skin. An outer layer of cotton is typically added to improve appearance because, in Mr. Parke's view, it had a more attractive weave.

Artificial-Limb Design Parameters

Mr. Parke observed that it would be inefficient to have a bank clerk wearing an arm that has been designed for a lumberjack: "A lumberjack needs an arm of maximum strength, and this would be unnecessarily heavy for a man who lifts nothing more than a stack of dollar bills." Therefore, his company wanted to know the amputee's occupation. Knowing this, appropriate materials and a hook designed for the amputee's occupation could be selected.

Mr. Parke was asked what he thought were some important parameters an engineer should consider in prosthetic design. "A lot of engineers try to over-design. I have met engineers who can design artificial arms that are almost perfect from a functional standpoint, but they are so complicated they would fill up an entire room. The size of such an arm would be just one of its problems. Amputees have what we call a 'gadget tolerance.' That is, there is a limit to the number of clever devices you can put in a prosthesis. If a guy has to wiggle his ears to move his hook, he will refuse to wear one.

"Reliability is a very important design consideration," Mr. Parke continued. "Imagine a situation like this: An amputee who needs his arm for his work is living 100 miles from his prosthesis-fitting facility, and a 50¢ spring breaks in his elbow joint. He is unable to use the arm, so he has to leave his job to get it repaired. This could take a number of weeks, since the fitting facilities normally do not carry spare parts.

"We tell amputees that they should treat their artificial arms as they would treat real ones. Unfortunately, we can't always count on their doing this. I heard about a mechanic, for example, who would stick a crowbar under a car to lift it up. After he had gotten it up, he would wedge his split hook in where the crowbar was and hold the car with his hook while he moved the crowbar to a new position. That guy would need a heavy-duty-steel split hook."

Mr. Turner had examined units after several years of use. "Once in a while, we get back an elbow that is bent or broken. I have tested these elbows myself, and I know that it takes a lot of force to break one. They still seem to do it. I got an elbow from Florida that had sand and salt water in it. This is one of the reasons we make our metal parts of anodized aluminum and stainless steel. However, the anodizing can wear through after a while, and even sweat is corrosive enough to attack unprotected aluminum. Despite these problems with a few units, our arms usually can be worn for a number of years before they require maintenance. Most of the bearing surfaces, for example, use Oilite bushings and rarely need to be lubricated.

"The range in size of the people we must fit has caused some design problems," said Mr. Parke. "We scale our parts to fit several size ranges, and this means we need a lot of tooling for parts that would be made in small quantities even if there were only one standard size. Scaling causes some more problems. It can be difficult to build such things as springs in a size small enough for children. Also, it is difficult to assemble child-size mechanisms."

Problems with Existing Prostheses

"Our products have a record of continuous satisfactory performance," said Mr. Parke, "but there are some areas where they could be improved. We need a better constant-friction wrist connection. It is desirable to be able to rotate the hand to different positions about the longitudinal axis of the forearm. A welder might want to rotate his hand to one position for holding a welding rod and to another position for picking up a fork. The rotating connection should have sufficient friction so it will stay in the position to which it is set, but it should not have so much friction that it becomes difficult to adjust. Our present connections tend to bind or loosen up, and are continually being readjusted. The cable that opens the split hook exerts a torque on the hook when it is pulled, so if the wrist joint is loose, the hand will turn when you try to open it. The ideal joint should have some provision for setting the friction initially, so it would be suitable for either a logger or a baby, and it should be capable of being repositioned about 80 times a day for the life of the arm. An arm can

last as long as 20 yr, but the average life is from 3 to 5 yr.

"Our elbow-locking mechanisms could also be improved. One problem is that when an amputee wants to lock or unlock his elbow, he has to make an awkward 'monkey motion' with his shoulder; he moves his stump down and back while his shoulder moves forward. This is a difficult motion for amputees to learn. The best way of solving this problem would be to design a solenoid elbow lock that could be controlled by flexing a muscle against a sensitive electric switch. Two federal agencies are presently working on such an electric elbow, but they haven't come up with anything practical yet. One of their designs does not permit the arm to swing free when the elbow is unlocked. With this design, an amputee would look stiff if he walked with his arms at his sides.

"A second problem is that, due to a lack of available space, some of our elbow-locking mechanisms are not very strong. We usually mount the locking mechanisms inside the elbow, and in that location we have sufficient room to make the unit sturdy enough for most applications. Unfortunately, some amputees have had amputations at the elbow joint,* and their stumps fill up the space where we would normally locate the locking mechanism. In such a situation the most efficient thing to do would be to amputate the arm back to the place that would permit us to use the internal elbow unit. However, it's hard to make a person go back to the hospital after he has once had an amputation. Also, insurance companies pay for amputations by the inch, so they are reluctant to have any arm cut away for prosthetic-fitting purposes. We are forced to design around this problem and mount a locking mechanism along the outside of the prosthesis. We can't let the mechanism protrude very far from the sides of the elbow, since it would be cumbersome, so we must make it smaller than we would like. As a result, our externally mounted unit can only take half the torque (600 in.-lb) of the internal unit. We have a real need for a stronger external unit." (The existing mechanism is described in Exhibit 5 — J. L. Adams Corporation.)

"We have another problem that is caused by the lack of space in elbow disarticulation prostheses. There is insufficient room to mount a joint that will allow the forearm to turn about the longitudinal axis of the upper arm. The stump occupies the space where the joint (usually referred to at Adams as a 'turntable') is usually located. In this situation we must fix the angle of the arm, and there is no way of readjusting it.

"Appearance is a problem with our counterbalance unit, which is used to support the weight of the forearm [Exhibit 6 — J. L. Adams Corporation]. We don't even make it for the elbow disarticulation arm

---

* Amputation at a joint is known as "disarticulation."

because there is no room for it inside the joint and it protrudes and makes the arm look too awkward. Ideally it should not protrude, and yet it must be adjustable in force to counterbalance terminal devices of different weights. The maximum weight of a cosmetic hand is 1 lb, and the maximum weight of a hook 6 oz. A forearm itself typically weighs roughly 1 oz per longitudinal in.

"We can find lots of shortcomings like these in the existing product line where I suppose application of engineering talent could help us. On the other hand, we could also expend that kind of talent on new developments like those of the Academy of Sciences list. Our competitors presumably face similar choices. We feel it is important to lead them in quality, but at the same time we have to remember that our present strong position in the market, in fact the existence of this company, was made possible by pioneering in new concepts. Unfortunately, we are too small to operate on all technological fronts at once. We have to aim our modest engineering capacity carefully."

Discussion Questions

1   Describe the qualifications this company should seek if it were going to hire a full-time engineer.
2   If you were hired as an engineer by the company, what would be your plan of action? How might a more experienced person be able to operate differently?
3   Select a design project for the J. L. Adams Company and carry it forward as well as you are able.

Exhibit 1 — J. L. Adams Corporation List of prosthetics design problems*

INTRODUCTION

This brochure has been prepared by the Committee on Prosthetics Research and Development to serve as a guide to the Faculty of Engineering and Design Schools in developing problems of the handicapped for student design projects.

Listed herein are some typical design problems relating to Prosthetics and Orthotics together with a brief description. In addition, a few specific problems have been developed in more detail to illustrate how they might be used in student programs at both under-graduate and graduate levels.

It is hoped that student participation in this field will stimulate new concepts in the solution of the problems, provide the students with very real projects, and interest them in Rehabilitation as a career.

Further information may be obtained from:

> Mr. A. Bennett Wilson, Jr.
> Technical Director
> Committee on Prosthetics Research and Development
> 2101 Constitution Avenue
> Washington 25, D. C.

CONVERTING ELECTRICAL ENERGY TO PNEUMATIC ENERGY

A number of reasonably well-designed components now exist for use of external power in upper-extremity prosthetic systems. Development work so far has yielded pneumatic motor mechanisms of high quality and reliability. However, employment of pneumatic systems requires rather bulky yet portable energy sources (small compressed CO_2 containers or "bottles") which are charged at some inconvenience to the wearer using large cylinders to be stored in his home. A.C. electrical sources are available in most homes and offices; ideally electrical energy should thus be used to charge a portable pneumatic power source. There is a need for development of a simple portable system by which A.C. power can be used to charge a portable pneumatic power source.

References:

There has been some work done by Heather et al., in Delaware, on a system of this sort, but there is very little in the form of reference material except materials which deal with pneumatic motor mechanisms already developed. Some pertinent information regarding end use in — The Application of External Power in Prosthetics and Orthotics, National Academy of Sciences, National Research Council, Publication 874, 1961, Washington, D. C.

COMPRESSED GAS GENERATOR FOR POWERED PROSTHESES

Gas-operated, externally powered upper extremity prosthesis components require frequent refilling of gas containers at pressures of 1500 to 2000 psi. by means of special filling devices. Substantial improvement could be effected if the tanks could be filled by solid pellets, liquids or powdered chemicals at room temperature which would produce adequate volumes of gas inside the tank under the action of a catalytic additive. In effect, the user would simply bleed the depleted tank to atmospheric pressure, open it, deposit measured amounts of the appropriate materials, seal the tank and wait for the reaction to produce the gas. Ideally, the reaction should be controlled by a pressure threshold and limit.

---

* Excerpts from a brochure prepared by the Committee on Prosthetics Research and Development of the National Academy of Science, National Research Council.

Reference:

No specific reference available. Some pertinent information regarding end use in — The Application of External Power in Prosthetics and Orthotics, National Academy of Sciences, National Research Council, Publication 874, 1961, Washington, D. C.

DESIGN OF AN EXTERNALLY POWERED ARM PROSTHESIS

A person has been involved in an accident resulting in the amputation of one arm through the shoulder joint. This type of amputation, known as a shoulder disarticulation, is one of the most difficult to fit with a conventional "body powered" arm prosthesis. The lack of an upper arm stump as an active source of power for movement and control of the prosthesis requires that an external source of energy be provided. This source of energy must be small enough and lightweight enough to be carried by the amputee under his clothing. The arm should provide a replacement for those functions considered essential in activities of daily living. A considerable body of knowledge is available which will be helpful in establishing design criteria for such a device.

References:

1. Artificial Limbs, May 1955 Issue.
2. Artificial Limbs, Sept. 1955 Issue.
3. Human Limbs and Their Substitutes, Klopsteg, Wilson et al. An account of the research carried out under the Committee on Prosthetics Research and Development, National Academy of Sciences, through 1954.
4. Snelson and Marsh, "Design and Development of an Externally Powered Arm Prosthesis." An acount of a similar development.
5. Paul and Mann, ASME Paper 62-Wa-121, "Evaluation of Energy and Power Requirements for Externally Powered Upper-Extremity Prosthetic and Orthotic Devices."
6. "The application of External Power in Prosthetics and Orthotics." Published 874, National Academy of Sciences, National Research Council.
7. University of California, Dept. of Mechanical Engineering, ME 124 — Design Project — Spring 1964.
8. Keller, A. D., C. L. Taylor, and V. Zahm, Studies to determine the functional requirements for hand and arm prosthesis, Department of Engineering, University of California, Los Angeles, 1947.

TERMINAL DEVICE (T. D.) SHAPES

Most T. D.'s are based on the split hook or two fingered hand principle. A study of T. D. shapes to provide optimum grasp for a maximum number of tasks is indicated. The results of this study would have application in remote manipulators as well as prosthetics.

References:

1. Northrop Aircraft, Final Report on Artificial Arm and Leg Research and Development, 1951 under National Research Council.
2. Orthopaedic Appliance Atlas Volume II, published by J. W. Edwards, Ann Arbor, Michigan.
3. Studies of the Upper-Extremity Amputee, Prosthetic Usefulness and Wearer Performance, by Hector W. Kay and Edward Peizer, Article in Artificial Limbs, Autumn 1958, National Research Council.
4. Keller, A. D., C. L. Taylor, and V. Zahm, Studies to determine the functional requirements for hand and arm prosthesis, Department of Engineering, University of California, Los Angeles, 1947.

OPTIMUM FRICTION AND RESILIENCE
FOR ARTIFICIAL HANDS AND HOOKS

The abilities to grasp and release objects are particularly a function of the friction characteristics of the materials in contact and the resilience of the grasping surface. Although high friction is desired for holding — low friction has advantages in sliding hands in the pockets, etc. The optimum values for friction and resilience for various areas of the appliance should be determined.

References:

1. Report Second Workshop Panel on Upper-Extremity Components, Subcommittee on Design and Development, COMMITTEE ON PROSTHETICS RESEARCH AND DEVELOPMENT, National Research Council, Washington 25, D. C.
2. Studies of the Upper-Extremity Amputee, Prosthetic Usefulness and Wearer Performance, Hector W. Kay, Edward Peizer, in Artificial Limbs, Autumn 1958, National Academy of Sciences, National Research Council.

HAND AND HOOK: COMBINED MECHANISM
FOR VOLUNTARY-OPENING AND VOLUNTARY-CLOSING

Several artificial hands and hooks have been designed using different types of internal mechanisms so that through voluntary cable motions amputees could either *open* or *close* the hand against a spring force used to return the hand to the original position. Hands or hooks have been made with either function available but not both. It would be desirable to design a hand or hook which with very simple modifications could be converted from one type of function to the other.

References:

1. A voluntary-opening and voluntary-closing hand or hook can be made available as well as drawings showing the mechanisms of both types.
2. Descriptions of artificial hands and hooks can be found in Orthopaedic Appliance Atlas, Volume II, published by J. W. Edwards, Ann Arbor, Michigan, 1960 Issue.

SOCKET VENTILATION BY NORMAL MOTION

Recent designs of sockets for the amputee's stump have employed very close fitting non-porous plastics. Some amputees have complained of perspiration accumulation and discomfort from lack of ventilation. Socket ventilation should enable the stump to "breathe" and to allow perspiration and water vapor to escape. A system design is needed particularly for the lower extremity prosthesis wherein certain functions such as the required resistances to foot flexion and extension or knee flexion and extension could be used to provide ventilation around the stump surface. Many above knee sockets use suction to keep the limb on. It is desirable that the ventilation system be applicable to suction sockets.

References:

There is little reference material available except that which discusses the desired functions of knee and foot-ankle mechanisms which might be used for ventilation sources. One approach to socket ventilation is described and has been developed by the Navy Prosthetic Research Laboratory, U.S. Naval Hospital, Oakland 14, California.

CENTER CONTROL TERMINAL DEVICE

Artificial hands or hooks are usually operated by a control cable connected to shoulder harness — to improve appearance and to enhance efficiency at various position of wrist rotation, the appliances should be developed with a central or axially located control cable or rod.

References:

Northrop Aircraft Inc., Final Report on Artificial Arm and Leg Research and Development.

Second Workshop Panel on Upper-Extremity Components, Subcommittee on Design and Development, COMMITTEE ON PROSTHETICS RESEARCH AND DEVELOPMENT, National Research Council, Washington 25, D. C.

CO_2 PORTABLE BOTTLE RECHARGING SYSTEMS

Carbon Dioxide stored in small bottles is used as an energy source for activating artificial arms and upper extremity bracing. Refilling the bottles from a larger container is a hazardous task that is difficult to perform. There is a need for a home refilling method that is safe and can be operated easily by a handicapped or one-armed person.

References:

The Application of External Power in Prosthetics and Orthotics, National Academy of Sciences, National Research Council, Publication 874, 1961, Washington, D. C.

ELECTRICALLY OPERATED–PNEUMATIC VALVES

Many prosthetic and orthotic devices use CO_2 stored in small bottles and regulated to about 75 psi to operate pistons etc. to provide active function. Also under development are electronic signal and control systems (EMG for example), and it is desirable that these controls be applied to the pneumatic actuators. There exists a need for an electric-pneumatic valve whereby a variable electrical signal can be used to control a correspondingly variable gas flow. A 12 volt electrical system can be considered standard.

References:

The Application of External Power in Prosthetics and Orthotics, National Academy of Sciences, National Research Council, Publication 874, 1961, Washington, D. C.

Exhibit 2 — J. L. Adams Corporation

Exhibit 3 — J. L. Adams Corporation

* WRIST DISARTICULATION UNIT

All stainless steel, friction type—used on very long below elbow stumps and wrist disarticulations. Made as thin as possible to save minimum overall length.

Until introduction of * Oval Wrist listed below, this unit was used almost universally for this type amputation.

WA-600 Wrist Disarticulation Unit, adult size.
Solid stainless steel.
2" overall outside diameter.

WA-600A Wrist Disarticulation Unit, child size.
Solid stainless steel.
1 3/8" overall outside diameter.

* OVAL FRICTION WRIST

Prosthetist and amputee alike will be pleased with the fine appearance and function of the completed prosthesis when using this new oval shaped wrist.

Developed for improved cosmetic appearance. Particularly suitable for wrist disarticulation or very long B.E. stumps having conical shapes. Also being used on other applications where oval shape is desirable.

OW-100 Adult size, aluminum with stainless threads.
OW-75 Child size

* ADJUSTABLE FRICTION WRIST

Cap screws on to hold stainless steel threaded plate against base section. This threaded plate can be positioned in the most desirable location for proper friction of terminal device—turn plate one notch for new adjustment. Threaded plate can be replaced if threads become damaged. Liked by Prosthetists because of large opening for breakout and wax drainage.

AF-700 Adjustable Friction Wrist
Aluminum with stainless threads.
2" overall outside diameter.

FM WRIST DISCONNECT
(Quick-change type)

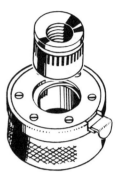

Developed under the program directed by the National Academy of Sciences Advisory Committee on Artificial Limbs, this wrist unit is particularly useful where both hand and hook are used alternately.

Light pressure on the control button maintains the terminal device in the arm, but allows it to be manually pronated or supinated. Pressure on the terminal device, towards arm, re-engages lock. Heavy pressure on the control button ejects the terminal device from arm.

Two inserts, FM-104, are furnished with each wrist. 1/2-20 thread standard, 1/2-27 on request.

FM-100 Wrist Disconnect
Complete with two FM-104 inserts.
2" overall outside diameter.

* QUICK-CHANGE WRIST

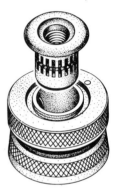

A positive locking quick-change wrist disconnect. Enables the amputee to position or interchange from hand to hook in a matter of seconds.

Wrist face is rotated in one direction to allow removal of terminal device. By rotating in opposite direction, terminal device is in "free wheeling" and can be manually rotated for correct position. With wrist face in central position, terminal device is in positive locked position and will not turn or pull out of wrist.

Some difficulty is experienced in rotating face when covered by cosmetic glove. Therefore, the FM-100 wrist shown on page 12 is preferred under these conditions.

WD-400 Standard weight
complete with two DA-101 inserts.
2" overall outside diameter.

WD-400S Heavy duty model
complete with two DA-101S inserts.
2" overall outside diameter.

* ECONOMY FRICTION WRIST

A simple screw-in friction wrist that is very popular with both Prosthetist and amputee. Available in a variety of sizes. Has stainless steel threaded bushing pressed into aluminum body.

1/2-20 thread standard, 1/2-27 thread on request.

Adult size available in solid stainless steel for heavy duty applications

STANDARD SIZE

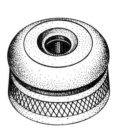

WE-500 Aluminum with stainless threads.
2" overall outside diameter.

WE-500S Heavy duty solid stainless steel.
2" overall outside diameter.

MEDIUM-SMALL SIZE

WE-300 Aluminum with stainless threads.
2" overall outside diameter.

CHILD SIZE

WE-200 Aluminum with stainless threads.
2" overall outside diameter.

INFANT SIZE

WE-100 Aluminum body threaded for Infant Mit.
Removable 1/2-20 thread stainless insert has
1/4-28 threaded hole for * #12P Infant
Hook. Insert can be adjusted for proper hook
position.

Exhibit 4 — J. L. Adams Corporation

1

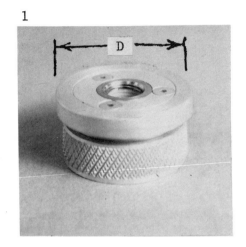

2

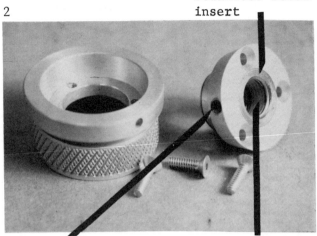

Stainless steel
insert

Friction adjustment
screw

Delrin plastic
insert

Picture 1 shows a typical mechanical wrist unit. This wrist holds a terminal device and allows it to rotate about the longitudinal axis of the forearm. Rotation is accomplished by manually screwing and unscrewing the terminal device.

The lower half of this unit is laminated to the plastic forearm. The upper half protrudes from the end of the forearm. Its top surface contains a threaded hole in which the bottom of the terminal device is screwed. The ends of the threaded section are made from stainless steel inserts, and the center section is a piece of Delrin plastic. Plastic is used to provide fairly constant friction when the terminal device is being turned. The steel inserts hold the plastic in place and prevent cross-threading of the plastic when a terminal device is being screwed in.

Picture 2 shows a partially disassembled wrist. The part on the right, which contains the threaded plastic insert, is removable so that it can be replaced if the plastic is damaged. A set screw, shown in this picture, pushes the threaded plastic section against the threaded part of the terminal device. It is used to adjust the amount of friction in the threaded joint.

The wrist is available in four sizes: Adult (D = 2″), Medium (D = 1-5/8″), Medium Small (D = 1-3/8″), and Small (D = 1-1/4″). It retails for about $12.50.

Exhibit 5 — J. L. Adams Corporation External elbow locking

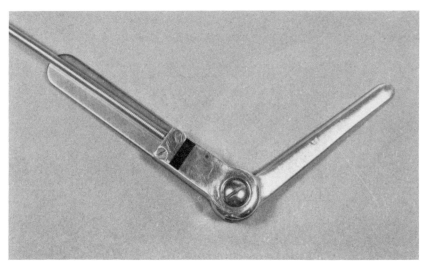

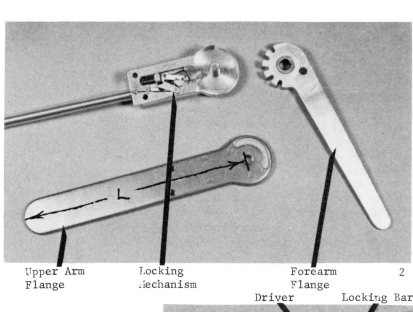

Upper Arm   Locking   Forearm   2
Flange   Mechanism   Flange

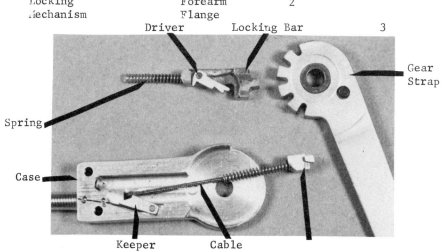

Driver   Locking Bar   3

Gear
Strap

Spring

Case

Keeper   Cable

The external elbow locking unit is used on elbow disarticulation prostheses to allow the amputee to lock his forearm in several positions (Child size: 5 positions; Medium size: 6 positions; Adult size: 7 positions), equally spaced through an angle of 145° about an axis that approximates the axis of rotation in the human elbow.

The forearm and upper arm flange are laminated into the plastic upper arm and plastic forearm respectively. Before lamination, the fitting facilities drill holes in these flanges so that the plastic will lock to the metal.

A partially disassembled unit and a prosthesis using one of the units is shown. Usually, one side of the elbow joint contains such a locking unit, and the other side contains a simple hinge. In cases where the amputee needs a heavy duty prosthesis, two of the external locking units are used. The elbow is made in three sizes: Standard (L = 6-1/2″), Medium (L = 5-1/2″), and Child (L = 4″). About 150 units are sold per year at a retail price of approximately $90.

We also see the disassembled locking mechanism.

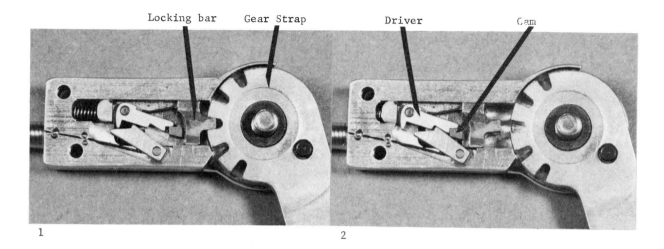

1   2

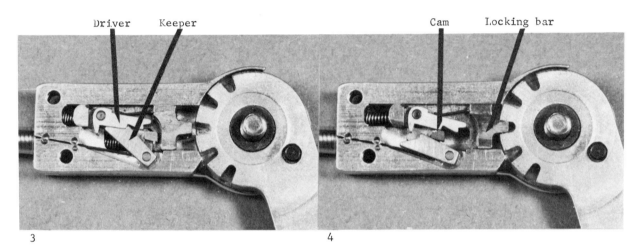

3   4

The operating sequence of the elbow locking mechanism is as follows. In picture 1, the locking bar is engaged with the gear strap and the elbow is locked. In picture 2, the cable has been pulled by the amputee, and the cam engages the driver, pushing the locking bar away from the gear strap.

In picture 3, the cable tension has been released, and the driver is engaged with the keeper. The elbow is now unlocked. Finally, in picture 4, the cable is pulled again, and the cam pushes against the keeper causing it to disengage from the driver. Then the coil spring connected to the locking bar pushes the bar back into one of the slots in the gear strap, thus locking the elbow.

Exhibit 6 — J. L. Adams Corporation   Forearm counterbalance unit

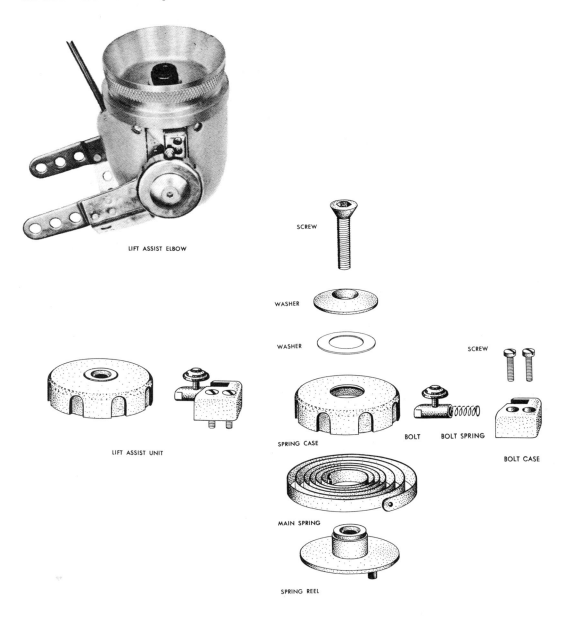

LIFT ASSIST ELBOW

SCREW

WASHER

WASHER

SCREW

LIFT ASSIST UNIT

SPRING CASE

BOLT   BOLT SPRING

BOLT CASE

MAIN SPRING

SPRING REEL

The forearm counterbalance ("lift assist") unit aids the amputee in positioning his forearm by counterbalancing the torque that the forearm exerts about the elbow joint. About 500 units are sold yearly as an accessory attached to the internal locking elbow. Retail price is about $25.

Spring counterbalance tension is set by manually rotating the spring case. The case is held in position by a sliding bolt. To release spring tension, the bolt is disengaged from the spring case by pulling on it.

JACK WIREMAN

Failure of a Ball Bearing

Part 2

The bearing company, when told details of the failure, specified a different bearing. Task tried it and it failed sooner than the first. Now what?

Following the failure of the 204SST5 bearing, Mr. Wireman used Orbit catalog data to calculate the manufacturer's design life of the bearing. According to his calculations, the bearing had a "B-10" life of 6,500 hr. He then sent the failed bearing to Orbit for examination. Orbit replied with the recommendation that a bearing with higher load capacity but identical external dimensions, either number M204BJHX2 or 204HJB1519, be used. Mr. Wireman decided to use the 204HJB1519.

The conclusion of the Orbit Corporation upon examination of the 204SST5 bearings from the first motor was that failure in the front bearings had been due to poor lubricity between balls and races. This opinion was expressed in a letter from Orbit of September 11, 1963, which is shown as Exhibit 1 — Jack Wireman (Part 2). The answer, Orbit asserted, was to use a bearing of higher load capacity and longer design life. Of the two bearings suggested by Orbit, Mr. Wireman chose the 204HJB1519 because it was much cheaper, costing only a few cents more than the previous 204SST5 bearing. The 204HJB1519 had a higher initial contact angle* (18° instead of 14°) and more balls (11 balls instead of 8) than the previous front bearing, and a design life of 19,000 hr, according to Mr. Wireman's calculations.†

Mr. Wireman was puzzled about why it should be necessary to use a bearing with a design life so much higher than the required operational life, but somehow the idea of using a heavier bearing seemed rea-

sonable. Experience with another motor also supported this approach. The other motor was an air-cooled version, which attached in the same way to run the same pump. It differed in that although its front-shaft bearing was fully immersed, the oil-flow path was around, not through, the bearing; and there was a seal that kept the hydraulic fluid out of the case, so the rotor and rear-shaft bearing were not immersed. The frame also differed in that it had cooling fins and allowed room for a larger front bearing. This was an Orbit 205 double-shielded, steel-retainer bearing, which had, according to Mr. Wireman's calculations, a lower contact angle but a design life of approximately 200,000 hr under the same loading conditions as the liquid-cooled pump motor. Another difference of the air-cooled motor was that it was used with MIL-5606 hydraulic fluid [a specification for which appears in Exhibit 2 — Jack Wireman (Part 2)], instead of Skydrol. No bearing failures had occurred on endurance tests of 2,500 hr with the air-cooled motor.

Since the use of a heavier-duty bearing seemed reasonable, and since the 204HJB1519 could be installed in place of the 204SST5 without otherwise altering the motor (the 205 bearing could not), thus limiting added costs, the decision was made to install it in all the liquid-cooled pump motors for Geyser Pump. Both qualification-test motors were fitted with the 204HJB1519 and 20 production motors that had been shipped were called back, torn down, and reassembled with the higher-capacity bearing, a process that consumed about 2 man-hr per motor. Drawings were modified to require the 204HJB1519 on all subsequent units.

There was considerable surprise and dismay when, after only 700 hr running on the reassembled, first qualification unit, the new bearing showed signs of failing. This motor had been put back on test to complete the original 2,500 hr. Since the first bearing had failed at 1,800 hr, there were 700 hr remaining

---

* The higher angular contact was produced by increasing the radial clearance in the bearing.
† The 1519 in the bearing number means diametrical allowance between balls was held between 0.0015 and 0.0019 in.

of the test. After running the 700 uneventful hours with the new bearing, the motor was stopped and disassembled to inspect those parts that had now completed 2,500 hr. The front bearing, which had as yet shown no symptoms of difficulty while running, was found to be just slightly spalled, signifying the onset of failure. Pictures of this bearing appear in Exhibit 3 — Jack Wireman (Part 2). A written description of the failed bearings is given in Exhibit 4 — Jack Wireman (Part 2). This description, dated December 9, 1963, was prepared by an independent consultant, Mr. Thomas Barish, a noted mechanical engineering consultant and author of over 20 published articles on bearings.

Discussion Questions

1 How did Jack Wireman define his problem? What definition should he now use?
2 How did the bearing company define the problem it needed to solve? What other definitions could it have adopted? Why did it pick the one it did?
3 How else could the motor have been designed so this problem would not have arisen? Should it have been different?
4 How else could the motor design be modified now? Develop some illustrative sketches. Should the motor design be changed now?

Exhibit 1 — Jack Wireman (Part 2)

ORBIT BEARING CORPORATION

September 11, 1963

Mr. Elmer Ward, Chief Engineer
Task Corporation
1009 East Vermont
Anaheim, California

Dear Mr. Ward:

This will summarize our recent telephone conversations and letters regarding the failure of an Orbit Precision 204SST5 bearing in a pump motor qualification unit and our recommendations to prevent recurrence of such failures. The pump runs at 6000 rpm in Skydrol at 150°F with a thrust load of 75 lbs. on the 204SST5 bearing and 20 lbs. on the opposed bearing, a 203SS5. The failure in the 204 size bearing was experienced at 1800 hours.

Examination of the 203SS5 bearing showed that it operated with normal contact angle in the presence of considerable contamination. There is also a very light running band where the bearing apparently operated under a reverse thrust condition. In spite of the evidence of contamination this bearing could have continued operating for a considerable time.

The 204SST5 bearing is a typical fatigue failure due to poor lubricity of the hydraulic fluid. In this type of failure a very fine surface spalling occurs on the inner ring raceway and erodes the metal until increased axial play causes failure by rubbing of the rotor and housing or increased loads due to the roughened raceway cause normal fatigue failure of the balls and outer ring. In this case both of these apparently happened. There is no evidence of inadequate alignment or poor mounting to aggravate the failure. It is doubtful that the contamination seen in the 203 size bearing had any significant effect. It is, however, possible that the use of a double shielded bearing helped keep debris in the ball path and accelerate the failure.

In order to increase the life of the bearing with Skydrol as lubricant and with the loads stated, it would be necessary to increase the capacity of the bearing. Because the failure at 1800 hours was a progressive type failure it would be necessary to make a substantial increase in capacity in order to prevent the application from being continually marginal in operation. The two alternates are first, a duplex pair of bearings which, as you say, would require considerable machine modification and, therefore, were not quoted; or a single bearing of different design for greater capacity. This greater capacity could be obtained by using the Orbit Precision M204BJHX2 or the 204HJB1519. Both of these bearings have the same envelope dimension as the 204SST5 originally used. The M204BJHX2 was developed particularly for use with heavy loads and low lubricity fluids. It has M50 tool steel rings and balls as shown on drawing SA-3711 which has been forwarded to you. The open, unshielded construction which permits an added flow of fluid which would help wash away the contamination or wear particles which might otherwise collect on the bearings. The oversize balls, high radial play, and high shoulders give the bearing a capacity slightly higher than that of the 205J bearing shown in the Orbit Catalogue. Based on the experience with the 204SST5 bearing there should be no problem meeting the desired life of 6000 hours with this bearing.

If, for economic reasons, the M204BJHX2 bearing cannot be used, the next recommendation is the Orbit Precision 204HJB1519 bearing. This bearing, because of its larger ball complement than the 204SST5, open construction, and high radial play, should give significantly longer life than the 204SST5.

The design life of the 204SST5 is 26,500 hours, for the 204HJB1519 it is

98,000 hours, and for the M204BJHX2 it is 235,000 hours. Assuming the failure at 1800 hours to be typical, this would give an expected life of 6600 hours for the 204HJB1519 and 16,000 hours for the M204BJHX2. This figure will be increased or the chances of attaining it will be much better with the use of the open, unshielded 204HJB1519 but it would be impossible to put a quantitative figure on this relationship without a considerable amount of testing. If the filtering system can be improved to effectively filter the fluid to a 10 micron level this should also have a beneficial effect on life.

In this discussion we have recommended use of bronze retainers based on your experience with this material in Skydrol. We have no field data on this and would normally have recommended comparative checks with phenolic and bronze. The bronze retainer has the added advantage of an extra ball in the complement, giving increased design life. The material used in these retainers is continuous cast Asarcon (80% copper, 10% tin, 10% lead) bronze. We have found it to be superior to all other types of bronze for use in high performance precision ball bearings.

Summing up, the failure of the 204SST5 bearing is a typical low lubricity fluid failure which we feel can be corrected to give good reliability at 6000 hours by use of the Orbit Precision 204HJB1519 bearing. It is further suggested that, in the qualification unit, the machine be disassembled at 3000 hours for bearing analysis at Orbit, reassembled, and continued for the full life of the bearing, with periodic checks.

Exhibit 2 — Jack Wireman (Part 2) Specification excerpts on MIL-5606 hydraulic fluid

REQUIREMENTS

3.1 Qualification. The fluid furnished under this specification shall be a product which has been tested and passed the qualification inspection specified herein, and has been listed on or approved for listing on the applicable qualified products list. Any change in the formulation of an approved product shall require requalification.

3.2 Materials. The fluid shall be clear and transparent consisting of petroleum products with additive materials to improve the viscosity-temperature characteristics, resistance to oxidation, and antiwear properties of the finished product.

3.3 Petroleum base stock requirements. The properties of the petroleum base stock used in compounding the finished fluid, before the addition of any other ingredients required herein, shall be as designated in Table I where tested as specified in 4.7.2.

Table I Properties of Petroleum Base Stock

PROPERTY	VALUE
Pour Point (max)[1]	−59.4°C. (−75.0°F.)
Flash Point (min)	93.3°C. (200.0°F.)
Acid or Base No. (max)	0.10
Color, ASTM Std (max)	No. 1

[1] Pour point depressant materials shall not be used.

3.3.1 Specific gravity. The specific gravity of the base stock shall be determined as specified in 4.7.2 but shall not be limited. Samples of base stock submitted for acceptance tests shall not vary by more than ±0.008 at 15.6/15.6°C (60.0°F) from the specific gravity of the original sample submitted for qualification tests.

3.4 Additive materials.

3.4.1 Viscosity-temperature coefficient improvers. Polymeric materials may be added to the base petroleum oil in quantities not to exceed 20 per cent by weight of active ingredient in order to adjust the viscosity of the finished fluid to the values specified in 3.5.

3.4.2 Oxidation inhibitors. Oxidation inhibitors shall be added to the base oil in quantities not to exceed 2 per cent by weight.

3.4.3 Antiwear agent. The hydraulic fluid shall contain 0.5 ± 0.1 per cent by weight of tricresyl phosphate, conformity to Specification TT-T-656.

3.5 Finished fluid. The properties of the finished fluid shall be as specified in Table II and 3.5.1 through 3.5.11.

Table II Properties of Finished Fluid

PROPERTY	VALUE
Viscosity in centistokes at 54.4°C. (130°F.) (min)	10.0
Viscosity in centistokes at −40°C. (−40°F.) (max)	500
Viscosity in centistokes at −54°C. (−65°F.) (max)	3000
Pour point (max)[1]	−59.4°C. (−75.0°F.)
Flash point (min)	93.3°C. (200.0°F.)
Acid or Base No. (max)	0.20

[1] Pour point depressant materials shall not be used.

3.5.1 Color. The fluid shall contain red dye in concentration not greater than 1 part of dye per 10,000 parts of oil by weight. There shall be no readily discernible difference in the color of the finished fluid and the standard color when tested as set forth in 4.7.3.

3.5.2 Corrosiveness and oxidation stability.

3.5.2.1 Corrosiveness. When tested as specified in 4.7.2, the change in weight of steel, aluminum alloy, magnesium alloy, and cadmium-plated steel subjected to the action of the hydraulic fluid shall be not greater than ±0.2 milligrams per square centimeter of surface. The change in weight of copper under the same conditions shall be no greater than ±0.6 milligram per square centimeter of surface. There shall be no pitting, etching, nor visible corrosion on the surface of the metals when viewed under magnification of 20 diameters. Any corrosion produced on the surface of the copper shall be not greater than No. 3 of the ASTM copper corrosion standards. A slight discoloration of the cadmium shall also be permitted.

3.5.2.2 Resistance to oxidation. When tested as specified in 4.7.2 the fluid shall not have changed more than −5 or +20 per cent from the original viscosity in centistokes at 54.4°C (130.0°F) after the oxidation-corrosion test. The acid or base number shall not have increased by more than 0.20 over the acid or base number of the original sample. There shall be no evidence of separation of insoluble materials nor gumming of the fluid.

3.5.3 Low temperature stability. When tested as specified in 4.7.2 for 72 hours at a temperature of -54 ± 1°C (-65 ± 2°F), the fluid shall show no evidence of gelling, crystallization, solidification, or separation of ingredients. Any turbidity shall be not greater than that shown by the turbidity standard.

3.5.4 Shear stability. When tested as specified in 4.7.4 the per cent viscosity decrease of the hydraulic fluid, measured in centistokes at 54.4°C (130.0°F) and at −40°C (−40°F), shall be no greater than the percentage viscosity decrease of the shear stability reference fluid nor shall the acid or base number have increased by more than 0.20 over the original acid or base number.

3.5.5 Swelling of synthetic rubber. When tested as specified in 4.7.2, the volume increase of the standard synthetic rubber L by the fluid shall be within the range of 10.0 to 28.0 per cent.

3.5.6 Evaporation. The residue after evaporation for 4 hours at 65.6 ± 3°C (150 ± 5°F) shall be oily and neither hard nor tacky when tested as specified in 4.7.2.

Exhibit 3 — Jack Wireman (Part 2)

Exhibit 4 — Jack Wireman (Part 2)

TASK CORPORATION　　　PUMP-MOTOR 56383

Examination of Failed Bearings:
(1) *First Set:* 1800 hrs. Pump End Brg. Orbit 204SST5
Inner Race: Completely failed all around circle, from over center at bottom to shoulder and shoulder rolled. Pitting rather shallow, no deep spots, and over all area. No signs of heating.

Race also showed one shallow band of contact near opposite shoulder, and narrow: may be rubbing of failed parts.

Too far gone to tell initial failure point. No difficulty in bore or clamped shoulders.
Outer Ring: Outer same but smeared over pitting, probably foreign matter pits from inner as surface not nearly so rough as inner. *Balls* also badly pitted, shallow type over nearly half of surface, with a few deeper pits or breaks from riding edge of inner after inner failed.

Measured one ball where not failed: .3125″ indicating no wear. (Shield rubbed on booster pump, after inner failed.)
Cage: 2 piece rivited bakelite compound with alum. side plates. Broken to disassemble brg. Heavy rubbing on bore of outer, one side only and severe pocket stress with many imbedded small steel flakes. Secondary failure: inner went first.
Conclusion: Heavy load, mostly thrust with possible contribution from lubricant of poor lubricity: (indicated by peculiar type of shallow pitting).
(2) Closed End Bearing: Orbit 203SST5

No failure: only surfaces badly spread with fine hard foreign matter pits. May have come from front bearing failure, but difficult to see how.
Inner Ring: Inner edge of contact fuzzy and measurements questionable. May have been two separate contact overlapped.

Bore had turned on shaft, rubbing mostly on both sides and not middle as the shaft were hollowed. Also shoulder rub and wear.
Outer Ring: Also pitted. O.D. rubbed and turned with most of contact at one end (outer end).
Balls and pressed steel cage very good.
(3) Second Set: 700 hours Pump End: Orbit 204HJB1519
Inner Ring: Shallow pitting load failure as above but caught much earlier. Extended only about 180° with tapering off at each end, and only for width of contact at max.
Outer Ring shows only small pitting from hard particles off inner. Otherwise in very good condition. No difficulty on bore, O.D. or shoulders of either race.
Cage (halo bronze type) extremely good in view of inner failure. No deterioration at all. Likewise balls good. No wear and only very faintly banded.
Closed End; Second Set: Orbit 203SST5 but shields removed.

Entire bearing in very good condition. No wear or loss of initial polish except slight greying of contact areas.
Inner showed fairly heavy load, per calculation. However contact may have been two contacts overlapped but still high and indicates more than 3 to 7 lbs. expected.

Complete absence of any pitting at all indicates that back bearing does not need shielding normally.

Thomas Barish
December 9, 1963

Four TECHNICAL ANALYSIS

How thick should the posts be for a 6-ft-high back-yard fence? Do-it-yourself homeowners constantly answer questions of this sort with no engineering training, and generally achieve satisfactory results. High-school hotrodders and surfboard shapers continually make design decisions with no formal analysis whatever, as do many independent inventors, some of whom are very successful. Ford, Edison, and Kettering made great technological contributions without college training in engineering. Much as engineering professors might prefer it some other way, the fact is that most practicing engineers, whether formally educated or not, make most of their engineering decisions without formal analysis. So why all the engineering-school emphasis on technical analyses?

How would you determine the amount of fuel a rocket would need to put a weather satellite in orbit, or predict what the orbit would be and how to get into it? Formal analysis is absolutely necessary for the solution to such a problem. How would you determine how thick to make a given dam or even how long a wire of what diameter to use in an electric heater? Cut-and-try might give workable solutions, but they would usually be vastly more expensive than those that can be produced with formal mathematical analysis. Such analysis can often predict things intuition would never guess, such as the fact that squeezing down the opening of a nozzle will increase speed-exit velocity of a gas only up to a certain point (the speed of sound). After that point the nozzle must diverge to produce further acceleration.

Sometimes physical arrangements are extremely difficult to treat analytically. Every year, for example, automobile companies smash hundreds of cars in crash tests to learn what they cannot predict analytically. Virtually all products, including electronic circuits — which have been designed with very heavy dependence on mathematical analysis — are given substantial testing to assure they will work as predicted. Even then, problems that were not revealed by either mathematical analysis or test occur later when used by purchasers. In the early fifties,

for example, the Cummins Diesel Company ran extensive tests on a diesel-powered racecar for the Indianapolis 500. It ran fine in the tests and did well in the race until, running in second place, it encountered the heavy buildup of rubber from tires of many cars on the track. The rubber dust plugged the air intake, and the car had to leave the race. One test to which it had not been subjected was running on a track full of other cars.

Testing is usually much more expensive than mathematical analysis. For that reason engineers generally prefer to make their preliminary decisions based on analysis, if only to reduce the number of required tests. The computation may be approximate, as when you estimate how much farther to go on a trip before refilling the gas tank, or exact, as when a slight variation in the efficiency of a steam power plant can cost thousands of dollars per year in fuel consumption. It may be fairly simple, involving simply application of a handbook formula, or complex, using expensive computer time to perform voluminous calculations.

Even the incredible numbers of calculations, however, will not assure that a solution is correct, because the underlying assumptions can still be wrong. Many Lockheed Electra airliners crashed — killing all aboard — in spite of extensive calculations and tests, until a particular vibration mode in the wing was discovered. There was virtually an infinite number of possible vibration modes, and all could never have been analyzed.

It is not the purpose of this chapter to explain how to perform mathematical analysis. Most other engineering courses do that. Instead, these cases aim to illustrate how the need for analytical treatment arises in practice, and the different ways engineers go about using pencil and paper in place of more expensive and time-consuming trial and error. Anyone can cut-and-try, but trained engineers can often do better.

A good place to start in applying analytical techniques is to draw a schematic picture (such as a free-body diagram or system boundary). Then list as many variables as you can imagine having something to do with the problem at hand, and wherever possible put applicable or desired numbers beside them. Then the trick is to find some relationship among the listed variables that will describe what physically happens. In most engineering lectures and textbooks this procedure is made straightforward by defining the course to fit the theory. Practice, on the other hand, is usually less orderly and requires mental groping. In "How to Solve It" by George Polya (Doubleday, 1957) he describes a realistic way to proceed:

> If you cannot solve the proposed problem try to solve first some related problem. Could you imagine a more accessible or related problem? A more general problem? A more special problem? An analogous problem? Could you solve a part of the problem? Keep only a part of the condition, drop the other part; how far is the unknown then determined, how can it vary? Could you derive something useful from the data? Could you think of other data appropriate to determine the unknown? Could you change the unknown or the data, or both if necessary, so that the new unknown and the new data are nearer to each other?

In the cases you will see engineers drawing from related problems in what they learned from earlier classes, from textbooks, and from handbooks. The resulting mathematical models always require certain assumptions that can totally invalidate the solutions if they are not correct. Consequently, it is always desirable to learn by what logic the models were derived. It is therefore a main aim of engineering education to develop the facility to trace out such logic.

Exercises

1 Pick some familiar object (slingshot, sprinkler, toaster, tetherball, and so on) and list variables that are important to how it works. Draw a schematic picture, and on it show the variables. Try to find some equations that describe its action. Identify possible reasons why they may not describe its action perfectly.

2 Sketch a dragster doing a "wheelie" and draw arrows for the forces acting upon it. What physical laws can you apply to it? How many different ways could you prevent the front wheels from leaving the ground?

3 Which would be harder to describe analytically, the forces imposed on a floor by a three-legged stool, or those by a carpet? What simplifying assumptions might you make in either case, and how do they affect your answer?

4 Describe the kinds of analytical models each of the engineers you have met so far in the cases would be best at constructing. If one of them were assigned to the job of another, tell how he could best compensate for weaknesses he might have in that new area.

5 Select one or more example problems from an engineering science course textbook. State the simplifying assumptions used in the model. Explain how you think the person who developed the equations in the model decided to make each of those particular assumptions. Which of the assumptions do you think might be the most serious in causing the model to depart from reality, and why? What could you do about it if you were designing something where the model might apply?

6 Name five things for which you think it would be extremely hard to develop effective mathematical models, and explain why you think so. What do you think would be the best that could be done in such circumstances if you had to predict performance?

7 For each of the following, name a variable you think would be valuable if predicted analytically. List other variables you think you would need to work into the model for predicting that critical variable. How would you find an appropriate model if you had to?
   a A steam boiler.
   b Supporting beams of a bowling alley.
   c An air conditioner.
   d A stereo speaker.
   e A gearbox.

IRV HOWARD

Mechanical Design in a Radar Simulator

What do buckshot and a mechanical engineer have to do with producing an electronic output? How well are they doing the job?

An electronic radar simulator being developed by the Simular Corporation, Los Angeles, California was to include a moving "window frame" 26-in. high by 50-in. wide, weighing over 150 lb. This frame, which was about 2-in. thick and stood in a vertical plane, had to be able to move 20 in. up and down and 20 in. from side to side (10 in. each way from center) past two optical scanning lenses, as schematically shown in Exhibit 1 — Irv Howard. To provide movement speed of 0.5 in./min, two servo motors of 4 watts each were geared to the frame through lead screws. For optimum performance of the servos it was necessary that they encounter approximately the same load in either direction of frame movement. Irv Howard, chief mechanical engineer at Simular, was concerned about the weight of the frame. In particular, he wanted to be sure that it would be adequately supported to be durable and hold the required tolerances, and that the servo motors would not be unduly loaded.

Basic Principles of the Simulator

The simulation process in the machine began with two optical lenses scanning two glass plates on which were drawn images of terrain the radar was to "see." The image on each plate (costing up to $1 million to produce; scale: 1 in. to 100 miles) looked somewhat like a photographic negative of the earth seen from high above. Each lens would look through the negative toward a light shining from the opposite side and producing a spot 0.001-in. in diameter. As the plate moved between lens and light in the dark and light regions of the negative, it would cause variations in the amount of light seen by the lens and correspondingly affect what appeared on the radar screen viewed by a pilot trainee sitting nearby in a simulated airplane cockpit. Depth of field for the lens to maintain resolution required that the distance between plate and lens be held within 0.005 in. of nominal.

When a radar views a real object it responds to two properties of it, one being the perpendicularity of its surface relative to the radar's line of sight, and the other being the reflectivity of the material from which the object is made. To impose these two types of effects the simulator used two plates, one for each effect, mounted side-by-side as in a stereo viewer, each with its own scanning lens and light. Signals from the two lenses were fed first to a computer, where they were coordinated, and then to the radar output and screen.

To operate properly the two pictures had to be positioned accurately relative to each other (within a tolerance of approximately 0.0001 in.). The size of each plate was to be 20 by 20 in., and each plate was to be located in a vertical plane (like a house window) between its lens and light source, these being 18 in. apart. In moving past the lenses to permit scanning, a maximum speed of 0.5 in./min was needed, and a translation accuracy in the plane of movement had to be held within 0.001 in. This made it important, in Mr. Howard's view, that no backlash be allowed in the mechanical system for moving the plate.

The general mechanical arrangement he expected would consist of a fixed frame plus two moving frames (see Exhibit 1 — Irv Howard). On the fixed frame would be a horizontal support shaft and a guide along which the horizontal frame would slide horizontally. On the horizontal frame would be a vertical support shaft and a guide along which the second moving frame containing the film plates could slide vertically. Thus the film plates could be moved both horizontally and vertically by electric servo motors acting through lead screws. He commented that no serious consideration had been given to laying the plates flat horizontally, partly because this would take more floor space for the machine and partly because the plates then might sag out of tolerance. Both these considerations in his view were important.

Background of the Project

The simulator project, which was now supported by a government contract calling for delivery of four machines, had begun in the proposal section of the company. A team of engineers, including Mr. Howard, had prepared a proposal in response to a government request for quotation. The expectation at Simular was that if the first four machines worked well, orders for others would follow.

During the initial discussions of the proposal team, Mr. Howard roughed out an overall sketch of the mechanical part of the machine, a copy of which appears in Exhibit 2 — Irv Howard. He expected, from previous experience on similar machines, that maintenance of tolerances would be important and that to lighten the servo-motor load it would be desirable to provide the moving frame with a counterbalance. For the latter he expected to leave free space at one end of the frame. However, to keep the floor-space requirement of the machine to a minimum, he decided it desirable to shape and mount the counterbalance as compactly as practically possible.

Mr. Howard, a graduate mechanical engineer, had been with Simular for 5 yr following 3 yr as a mechanical engineer in a large aircraft company. He commented that although he worked exclusively on mechanical engineering problems the company was primarily concerned with electronics. "There are plenty of mechanical problems on machines like this," he said, "but the mechanical engineer is seldom in charge. In an electronics company it's usually an electrical engineer who leads as project engineer. There are about 35 electrical engineers in the company and only 4 mechanical engineers.

"The mechanical engineer who does work like this has to be a generalist, ready to take on any sort of problem. One minute I'll be worrying about mechanisms and the next I may be on structures, shock and vibration, or air-conditioning problems. We've had specialists on subjects like stress analysis apply for jobs here, but have never hired any. The three graduate mechanical engineers under me have broad experience. They're good men because no matter what comes up, they can get the job done."

Mechanical Design

Mr. Howard's approach to the radar simulator design was first to sketch roughly the overall framework for mounting and moving the film plates, and then to concentrate on mounting them. Around both of the two 9-lb plates of 0.25-in.-thick glass he planned to fasten with rubber potting compound an aluminum support frame weighing 15 lb. This frame in turn would be fastened by thumbscrews into a larger frame of cast aluminum to run on ball bushings

(illustrated by catalog excerpt in Exhibit 3 — Irv Howard) along the vertical support shaft at one end with power from the vertical servo. A rail and roller guide at the opposite end of the frame would prevent rotation about the vertical shaft but would not provide any support. His estimate for the weight of the larger cast-aluminum frame and its bushings was 134 lb, so total weight with the two plates mounted would be 167 lb.

"To estimate the weight of the frame," he said, "I first drew it so it looked stiff enough. Then from the sketch I estimated the volume of metal and multiplied by its density. By spending more time I could probably have made the frame lighter. I could have made one, applied loads, measured deflections, and kept reducing the metal until it was no stiffer than it had to be and as light as possible. But I'm not a stress specialist, I'm a generalist. Also, this is not a mass production of the design. The main consideration was holding tolerances on the film plates. For the scanning lens to operate properly there should not be more than a half-thousandth play due to deflection, and this argues for making the frame and supports very rigid. Weight, on the other hand, is not so important in this machine, since it will be just sitting in a room, not flying in a plane. I have seen an engineer spend days refining for an ounce when a pound would do, and that, in my view, is a mistake."

Having roughed out the weight and dimensions of the frame, Mr. Howard thought he next should work on details of mounting it. In particular, he thought the next two components to concentrate on should be (1) the vertical support shaft; and (2) a counterbalance for the weight of the frame. His main question for the vertical shaft was what the cross-sectional dimensions should be. For the counterbalance he had decided to use a metal can filled with lead shot, hung by a cable over pulleys from the frame. The three questions here as he saw them were (1) what should be the dimensions of the counterweight can to hold the lead shot; (2) where should the counterweight cable be attached to the frame holding the film plates; and (3) what size cable should be used?

Evolution of the Counterweight Idea

Mr. Howard had thought of using a spring, but he quickly saw a difficulty in that the spring tension would vary as it was stretched, causing the counterbalance action to be too weak at one end of the travel and too strong at the other. He recalled that there was a type of spring called a negator, which exerts constant tension as it is stretched. But then, he reasoned, someone might lay the machine on its side, and there would be nothing to counteract the spring action. "Anyway," he said, "large negator springs are hard to get and hard to use. For this application it would have to be very large, and it would probably be dangerous to install. A thing like that could cut your fingers off.

"From the beginning of this design I assumed we would use a simple counterweight. I had needed one several years ago when we were using a little motor to move a heavy object vertically, and I've used them in a couple of designs since then, so the idea occurred easily. When I laid out the basic framework of this machine I left room at one end for the weight to hang."

A sketch showing the location of the rectangular space left open for the weight — approximately 6 by 7 in. horizontally and 52 in. vertically — appears in Exhibit 4 — Irv Howard. Mr. Howard considered the 7-in. dimension fixed because of mounting constraints of the lights and lenses. To make the weight fit compactly in the machine he assumed it would be rectangular except for a radius of 1.5 in. at one corner to fit the contour of 0.25-in.-thick adjacent aluminum shield. The 1.5 in. represented the minimum bend radius for such sheet, according to one of Mr. Howard's handbooks. The 6-in. dimension was, he thought, controllable, but to be minimized, since increasing it would enlarge the floor space required by the machine.

"As a material for the weight," he continued, "I wanted something to be heavy but not take up too much space. In my desk drawer I keep an issue of *Materials in Design Engineering* that lists specific weights of different metals. The heaviest is osmium, and below it are things like platinum and depleted uranium. The first reasonably priced material is lead, and that's what I picked.

"I thought of casting the lead in a bar, but since I wasn't sure precisely how heavy it should be I thought there must be some way to adjust the weight. So I considered using a stack of plates on a pole. It seemed simpler and more solid to make a can and pour in the right amount of melted lead. But again there was the problem of how to adjust the weight. And besides, hot lead can be poisonous. So I settled on the idea of filling a can with shot.

"Then there was the question of what volume to allow in the can. The magazine gave a specific weight for lead of 0.41 lb/in.^{3}, but none of my references said anything about the specific gravity of shot. I thought it would be lower for shot than for solid lead, and higher the smaller the ball size. But nobody even knew what sizes were available.

"So I drove a few blocks down to the Bruin Sport Shop and looked at shotgun-loading supplies. They had some shot of about a sixteenth of an inch in diameter. I imagined a close view of the shot piled up. I could see a sort of triangular prism of space between the balls. I just guessed that this would represent about 0.3 of the total volume, and so I assumed a 'space factor' of 0.7. In aircraft design

where weight is terribly important it would probably be worthwhile to study something like this analytically. But for this machine a guess seemed good enough. I don't particularly take pride in being a scientist, but I do in getting the job done."

For the can to hold the shot, Mr. Howard thought he would use steel sheet $\frac{1}{16}$-in. thick on the sides and $\frac{3}{8}$-in. thick on the top and bottom (0.287 lb/in.^{3}).

Mounting the Weight

As he envisaged it, the counterweight can would have a hole through it from top to bottom to allow for a vertical guide shaft along which the weight would run. He planned to select a splined shaft with a "ball spline," which would fasten to the counterweight can. To choose the shaft diameter he guessed a size and then computed the natural frequency of vibration the shaft would have when the weight was halfway between top and bottom. This frequency calculation appears in his notes reproduced as Exhibit 5 — Irv Howard. The formula used was one he had copied a few years earlier from a magazine article and written in the margin of a handbook.

"On a similar job some time back I learned the importance of making this calculation," he said. "That job involved a piece of equipment to be installed on shipboard, and the specifications required that it be tested on a (shake table) at frequencies ranging from 5 to 30 cps. We had used a much smaller guide shaft relative to the weight, and its natural frequency was somewhere in that range. When the shake table hit that frequency the weight started vibrating and threatened to break loose. It was terrible. There wasn't enough time to redesign and install a larger shaft, so we had to leave the counterweight out entirely. Luckily, the system had enough power to work all right without the counterweight. Experience like that has convinced me that generally it's better to overdesign than underdesign. So on this new machine I chose a natural frequency above 30 cycles even though no shake-table testing is required."

By making the guide shaft stiff enough to have a natural frequency of around 50 cps, Mr. Howard expected that there would be negligible deflection. Hence he made no analysis of bending strength or deflection. He thought sufficient clearance between the counterweight can and the stationary frame would be $\frac{1}{8}$ to $\frac{1}{4}$ in.

To suspend the counterweight he planned to use a steel cable and "shank ball," selected from a cable catalog, excerpts of which appear in Exhibit 6 — Irv Howard. "A safety factor of around 5 seems reasonable to me," he said. "That is a figure I've used before with success, and is in the range I would expect to find other engineers using. I think the figure first came to my attention in a magazine article I read some years ago, which said that for shipment by truck it was advisable to protect against a maximum loading of about 5 G's. That figure may also have appeared in some literature put out by shock-mount makers.

"But the cable should also be as flexible as possible, so I don't need an enormous pulley to run it over. Bending a heavy cable over a small pulley can make it fail in fatigue. So it's better not to use a cable heavier than necessary, and it's better to use one made of smaller strands.

"Now my questions are what dimensions to make the can, what size and type of cable to hang it with, and where to hook it onto the moving frame, whether off-center someplace or in the middle. I also need to choose the vertical support-bar diameter. Then I can design the frame for horizontal movement, after which come the horizontal supports and main frame."

Discussion Questions

1 Develop a critique of Irv's procedure thus far on this project.
2 Carry forward his job as far as you can with the information in the case.
3 What other information would you like to have if you were there, and how would you get it?

Exhibit 1 — Irv Howard

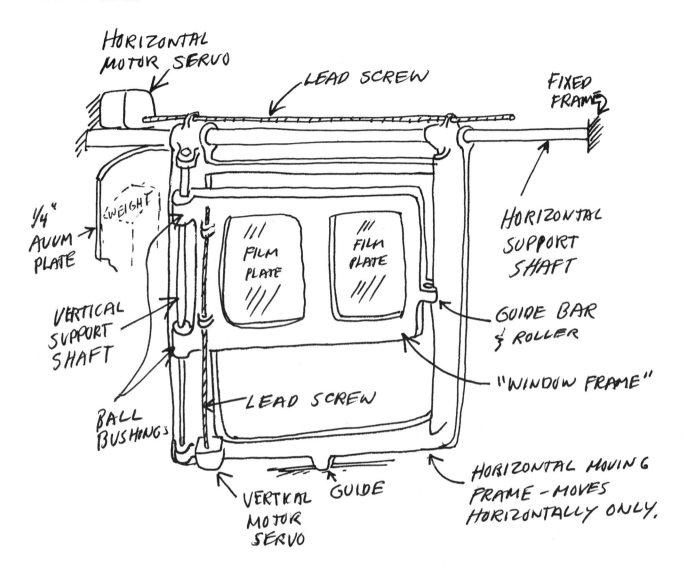

Exhibit 2 — Irv Howard

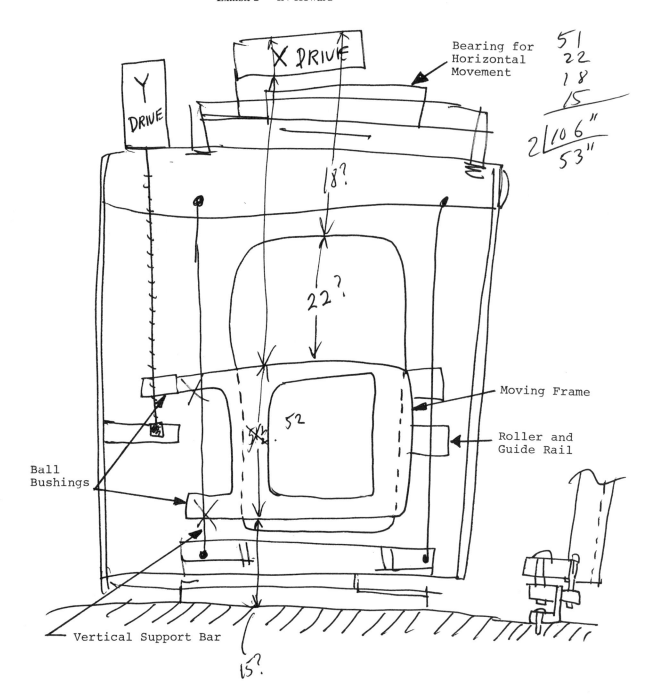

Exhibit 3 — Irv Howard

BALL BUSHING
ENGINEERING
SPECIFICATIONS

Table #2 — PRECISION Series A — Dimensions & Load Ratings

Series A Ball Bushing Number	Working Bore A (Inches)	Tol. +.0000 to	Outside Diameter B (Inches)	Tol. +.0000 to	Length C (Inches)	Tol. +.000 to	Dist. Between Retaining Rings D (Inches)	Tol.	Maximum Permissible Shaft Dia.*** (Inches)	Norm. Fit (Inches)	Tol. -.0000 to	Press Fit (Inches)	Tol. -.0000 to	Ball Diameter (Inches)	Number of Ball Circuits	Bushing Weight (Pounds)	Static (Pounds)	Rolling** (Pounds)	Series A Ball Bushing Number
A-4812	.2500	−.0005	.5000	−.0004	.750	−.015	.437	±.010	.2490	.5000	+.0005	.4990	+.0005	1/16″	3	.02	22	13	A-4812
A-61014	.3750	−.0005	.6250	−.0004	.875	−.015	.562	±.010	.3740	.6250	+.0005	.6240	+.0005	1/16″	4	.06	38	21	A-61014
A-81420	.5000	−.0005	.8750	−.0004	1.250	−.015	.875	±.010	.4990	.8750	+.0005	.8740	+.0005	3/32″	4	.08	72	46	A-81420
A-122026	.7500	−.0005	1.2500	−.0004	1.625	−.015	1.062	±.010	.7490	1.2500	+.0005	1.2490	+.0005	1/8″	5	.21	162	109	A-122026
A-162536	1.0000	−.0005	1.5625	−.0004	2.250	−.015	1.625	±.010	.9990	1.5625	+.0005	1.5615	+.0005	5/32″	5	.38	262	202	A-162536
A-203242	1.2500	−.0006	2.0000	−.0005	2.625	−.020	1.875	±.015	1.2490	2.0000	+.0010	1.9983	+.0010	3/16″	6	1.10	465	344	A-203242
A-243848	1.5000	−.0006	2.3750	−.0005	3.000	−.020	2.250	±.015	1.4989	2.3750	+.0010	2.3733	+.0010	7/32″	6	1.43	695	535	A-243848
A-324864	2.0000	−.0008	3.000	−.0006	4.000	−.020	3.000	±.015	1.9987	3.000	+.0010	2.9982	+.0010	1/4″	6	2.75	1100	850	A-324864
A-406080	2.5000	−.0010	3.7500	−.0008	5.000	−.025	3.750	±.015	2.4985	3.7500	+.0010	Not Normally Recommended		5/16″	6	5.50	1710	1380	A-406080
A-487296	3.0000	−.0012	4.5000	−.0010	6.000	−.030	4.500	±.015	2.9983	4.5000	+.0010			3/8″	6	9.50	2460	2000	A-487296
A-6496128	4.0000	−.0020	6.0000	−.0012	8.000	−.040	6.000	±.020	3.9976	6.0000	+.0010			1/2″	6	20.20	4400	3800	A-6496128

*Based on a shaft hardness of Rockwell 60C.   **Based on a travel life of 2 million inches. (See Page 25)
***For normal fit slightly larger shafts may be used with caution. (See Page 18)

STAINLESS STEEL: Series A, XA, ADJ and OPN BALL BUSHINGS are available made entirely of Stainless Steel. They are identified by the suffix SS following the part number (Example — XA-81420-SS). Series A and XA are stocked only in sizes up to and including 1″. Series ADJ and OPN are stocked in ½″, ¾″ and 1″. *For larger sizes see note on Page 29.*

Table #3 — SUPER PRECISION Series XA — Dimensions & Load Ratings

Series XA Ball Bushing Number	Working Bore A (Inches)	Tol. +.0000 to	Concentricity T.I.R. (Inches)	Outside Diameter B (Inches)	Tol. +.000 to	Length C (Inches)	Tol. +.000 to	Dist. Between Retaining Rings D (Inches)	Tol.	Maximum Permissible Shaft Dia.*** (Inches)	Norm. Fit† (Inches)	Tol. -.0000 to	Press Fit	Ball Diameter (Inches)	Number of Ball Circuits	Bushing Weight (Pounds)	Static (Pounds)	Rolling** (Pounds)	Series XA Ball Bushing Number
XA-4812	.2500	−.0003	.0005	.5000	−.0004	.750	−.015	.437	±.010	.2495	.5000	+.0005		1/16″	3	.02	22	13	XA-4812
XA-61014	.3750	−.0003	.0005	.6250	−.0004	.875	−.015	.562	±.010	.3745	.6250	+.0005		1/16″	4	.06	38	21	XA-61014
XA-81420	.5000	−.0003	.0005	.8750	−.0004	1.250	−.015	.875	±.010	.4995	.8750	+.0005		3/32″	4	.08	72	46	XA-81420
XA-122026	.7500	−.0003	.0005	1.2500	−.0004	1.625	−.015	1.062	±.010	.7495	1.2500	+.0005	NOT NORMALLY RECOMMENDED	1/8″	5	.21	162	109	XA-122026
XA-162536	1.0000	−.0003	.0005	1.5625	−.0004	2.250	−.015	1.625	±.010	.9995	1.5625	+.0005		5/32″	5	.38	262	202	XA-162536
XA-203242	1.2500	−.0004	.0010	2.0000	−.0005	2.625	−.020	1.875	±.015	1.2495	2.0000	+.0010		3/16″	6	1.10	465	344	XA-203242
XA-243848	1.5000	−.0004	.0010	2.3750	−.0005	3.000	−.020	2.250	±.015	1.4994	2.3750	+.0010		7/32″	6	1.43	695	535	XA-243848
XA-324864	2.0000	−.0004	.0010	3.0000	−.0006	4.000	−.020	3.000	±.015	1.9994	3.0000	+.0010		1/4″	6	2.75	1100	850	XA-324864
XA-406080	2.5000	−.0004	.0015	3.7500	−.0008	5.000	−.025	3.750	±.015	2.4993	3.7500	+.0010		5/16″	6	5.50	1710	1380	XA-406080
XA-487296	3.0000	−.0006	.0015	4.5000	−.0010	6.000	−.030	4.500	±.015	2.9992	4.5000	+.0010		3/8″	6	9.50	2460	2000	XA-487296
XA-6496128	4.0000	−.0010	.0020	6.0000	−.0012	8.000	−.040	6.000	±.020	3.9988	6.0000	+.0010		1/2″	6	20.20	4400	3800	XA-6496128

*Based on a shaft hardness of Rockwell 60C.   **Based on a travel life of 2 million inches. (See Page 25)   †For extreme precision, tolerance may be reduced.
***Slightly larger shafts may be used with caution. (See Page 18)

Table #4 — COMMERCIAL GRADE Series B — Dimensions & Load Ratings

Sold Only in lots of 250 or more.

Series B Ball Bushing Number	Working Bore A (Inches)	Tol. -.0000 to	Outside Diameter* B (Inches)	Tol. +.0000 to	Length C (Inches)	Tol. +.000 to	Dist. Between Retaining Rings D (Inches)	Tol.	Max. Perm. Shaft Dia. Norm. Fit (Inches)	Press Fit (Inches)	Housing Norm. Fit (Inches)	Tol. -.0000 to	Press Fit† (Inches)	Tol. -.0000 to	Ball Diameter (Inches)	Number of Ball Circuits	Bushing Weight (Pounds)	Static (Pounds)	Rolling*** (Pounds)	Series B Ball Bushing Number
B-4812	.2500	+.0020	.5000	−.0015	.750	−.020	.437	±.010	.2495	.2495	.5000	+.0010	.4980	+.0005	1/16″	3	.02	19	11	B-4812
B-61014	.3750	+.0020	.6250	−.0015	.875	−.020	.562	±.010	.3745	.3740	.6250	+.0010	.6230	+.0005	1/16″	4	.06	33	18	B-61014
B-81420	.5000	+.0020	.8750	−.0015	1.250	−.020	.875	±.010	.4995	.4990	.8750	+.0010	.8730	+.0005	3/32″	4	.08	61	39	B-81420
B-122026	.7500	+.0020	1.2500	−.0020	1.625	−.025	1.062	±.015	.7495	.7490	1.2500	+.0010	1.2475	+.0005	1/8″	5	.21	138	93	B-122026
B-162536	1.0000	+.0020	1.5625	−.0020	2.250	−.025	1.625	±.015	.9995	.9990	1.5625	+.0010	1.5600	+.0005	5/32″	5	.37	222	172	B-162536
B-203242	1.2500	+.0020	2.0000	−.0020	2.625	−.030	1.875	±.015	1.2495	1.2490	2.0000	+.0010	1.9970	+.0010	3/16″	6	1.10	400	292	B-203242
B-243848	1.5000	+.0020	2.3750	−.0020	3.000	.030	2.250	±.015	1.4994	1.4989	2.3750	+.0010	2.3720	+.0010	7/32″	6	1.40	590	455	B-243848

*Slight out-of-roundness may result from the heat treatment of Series B bearings, making it difficult to measure the true O.D. The bearing will return substantially to its original roundness when inserted into the recommended housing bore for either normal or press fit.   †Do not press fit in soft metal housings. Use normal fit to avoid shearing.
**Based on a shaft hardness of Rockwell 60C.   ***Based on a travel life of 2 million inches. (See Page 25)

COEFFICIENT of FRICTION

*Low friction is an important
feature of Ball Bushings.*

The coefficient of friction of BALL BUSHINGS — 0.001 to 0.004 — is extremely low, and is approximately that of *radial* ball bearings. It is far less than the coefficient of friction of sliding surfaces and, more important, *is far more constant.* This applies to the coefficient of rolling or operating friction as well as to the coefficient of static or break-away friction. From a practical standpoint, this low coefficient of friction is minute enough to be disregarded in most applications. A *constant* low coefficient of friction is imperative in many applications where stick-slip action impairs performance.

Extensive testing under controlled conditions has determined coefficients of rolling and static friction for the entire series of BALL BUSHINGS. These values may be used to estimate forces required to overcome frictional resistance in specific applications.

The formula used to determine frictional resistance during operation is:

$P = L \times f_r$, where P = Frictional resistance (pounds)
  L = Applied load (pounds), perpendicular to centerline of shaft
  f_r = Coefficient of rolling friction

The formula used to determine static *frictional resistance is:*

$P = L \times f_o$, where f_o = Coefficient of static friction

Following is a list of coefficients of rolling friction (fr) of BALL BUSHINGS operating on hardened and ground shafts of recommended diameters. These values are grouped according to the number of ball circuits in each Bushing as friction coefficients are constant among Bushings having 3 and 4 ball circuits but slightly less for Bushings with 5 or 6 ball circuits. To make the table easy to use Bushing sizes in each group are also listed.

The values for the coefficient of static or break-away friction are also listed. These values are not measurably affected by the number of ball circuits in the Bushing or by conditions of lubrication.

Variables affecting friction of Ball Bushings

SPEED — There are no appreciable variations in coefficients of rolling friction at various speeds.

LUBRICATION — Dry BALL BUSHINGS have the lowest coefficient of friction due to the complete absence of lubricant surface tension (meniscus drag) effects. Values for grease lubrication range from 100% greater in the smaller sizes to 20%-to-50% greater in the larger sizes. Oil lubrication (medium-heavy, viscosity 64 c.s. @ 100° F.) gives frictional values slightly higher than those for grease lubrication.

SEAL FRICTION — Where seals or flexible bellows are used to retain lubricant or to prevent entry of foreign particles, frictional resistance from these elements must also be taken into consideration in calculations made to determine total frictional drag. Where minimum friction is desired, sealing elements with minimum frictional drag are used or sealing elements are eliminated by design. BALL BUSHINGS are not as critical as radial ball bearings with respect to foreign particles as they are pushed aside by the ball action and are not trapped between balls and ball conforming grooves.

Coefficients of Rolling Friction (f_r) of Ball Bushings

$$f_r = \frac{P}{L}$$ where P equals *frictional resistance* and L equals *applied load*

BUSHING I.D.	NUMBER of BALL CIRCUITS	CONDITION of LUBRICATION	LOAD IN % OF ROLLING LOAD RATING (for 2,000,000 inches of travel life)				
			125%	100%	75%	50%	25%
¼″, ⅜″, ½″	3 & 4	No Lube	.0011	.0011	.0012	.0016	.0025
		Grease Lube	.0019	.0021	.0024	.0029	.0044
		Oil Lube	.0022	.0023	.0027	.0032	.0045
¾″, 1″	5	No Lube	.0011	.0011	.0012	.0015	.0022
		Grease Lube	.0018	.0019	.0021	.0024	.0033
		Oil Lube	.0020	.0021	.0023	.0027	.0036
1¼″ thru 4″	6	No Lube	.0011	.0011	.0012	.0014	.0019
		Grease Lube	.0016	.0016	.0017	.0018	.0022
		Oil Lube	.0018	.0018	.0019	.0021	.0027

Coefficients of Static Friction (f_o) of Ball Bushings

LOAD IN % OF ROLLING LOAD RATING				
125%	100%	75%	50%	25%
0028	.0030	.0033	.0036	.0040

Values are based on use of shafts of recommended diameters, hardened to Rockwell 58-63C.

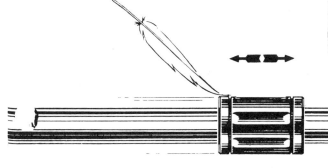

Exhibit 4 — Irv Howard

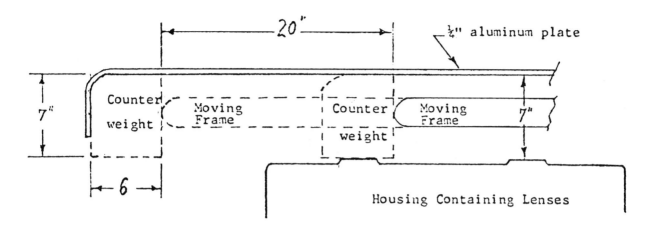

Top View

Positions of the frame and counterweight which move
both vertically and horizontally are shown in two
positions, one at the left limit of travel and the
other at the right limit of travel.

Exhibit 5 — Irv Howard

6/10/63

WT. OF VERT. MOVING BACONEY

(SEE 3/28, ETC)

```
2 PCS 20×20 GLASS ——————————— 18 LBS
2 FRAMES AROUND GLASS (STL) ——————— 15 LBS
2 BALL BUSHINGS ——————————— 11 LBS
ROLLERS & BKTS, ETC ——————— 11 LBS
MAIN CASTING FRAME (ALUM) —— 112 LBS
                                    ————————
                                    167 LBS
```

Amount to be Counterbalanced ——→ **167 LBS**

$$C' \ WT. \ VOL. = \left(\frac{167 \cdot 3}{.141 \ LB/IN} \times .7 \ \overset{SPACE}{FACTOR} \right) = 600 \ cu^3$$

Counterweight Volume

IF WE USE SPLINED SHAFT ——— 52" LONG, $170 \frac{LB}{CWT.}$

Counterweight Guide Shaft Frequency Determination

ASSUME 2" SHAFT — HAS 1.66 ROOT DIA.

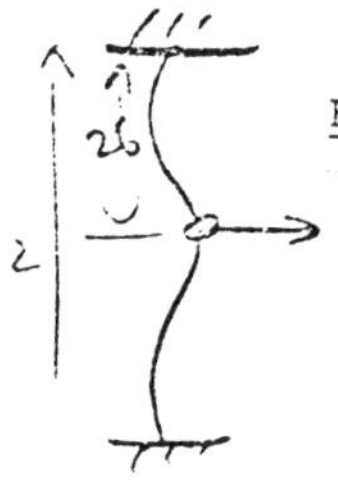

$$f_m = 43.3 \sqrt{\frac{3 \times 10^7 \times 5 \times 1.66^4}{170 \times 10^2 \times 52^3}} = 43.3 \sqrt{\frac{15 \times 1.66^4}{1.7 \times (5.2)^3}}$$

Formula From Magazine

$$= 43.3 \sqrt{.477} = (43.3)(.691) = 29.9 \ CPS, \ SAY \ 30 \ CPS$$

— HOWEVER —

← THIS IS TOO SIMPLE A REPRESENTATION —

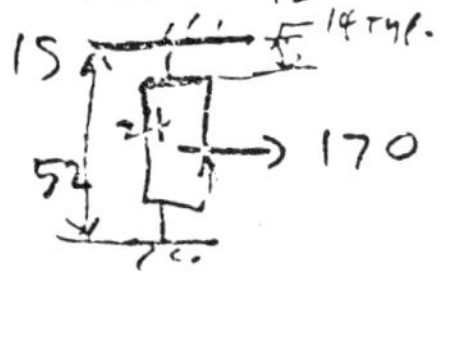

ACTUALLY
USE 2" SPLINE (MIN.)
USE 2½" SPLINE (IF POSS.)

C_{eff} for = 70, and 43.3

SO, we should have > 30 CPS

WITH $2\frac{1}{2}$ DIA. SPLINE $\left(ROO \rightarrow 2\frac{1}{2} = 2.1\right)$

$$f_m = 43.3 \sqrt{.477 \frac{(2.1)^4}{(1.66)}} = 43.3\sqrt{1.22} = 43.3 \times 1.105 = 47.8 \text{ cps}$$

(SIMPLIFIED) CALC.

MW.

IF WE HAVE LARGER COEFFIC. (WHICH WE DO) — WILL BE > 47.8 cps

coeff should be at least 50

giving $f_m = 50 \times 1.105 = 55.2$ cps

Make A $2\frac{1}{2}$ DIA. Spline — (TRY)

Exhibit 6 — Irv Howard

SPECIFICATIONS

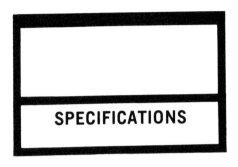

WIRE ROPE FOR ASSEMBLIES

Tables below cover Preformed galvanized steel and stainless steel most generally used for wire rope assemblies.

PREFORMED

Breaking Strength in POUNDS | **Breaking Strength in POUNDS**

DIAMETER IN INCHES	GALVANIZED STEEL			STAINLESS STEEL			DIAMETER IN INCHES
	MIL-C-6940	MIL-C-1511	MIL-C-1511	MIL-C-5693A	MIL-C-5424	MIL-C-5424	
	 1 x 19 ** ††	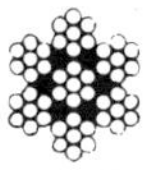 7 x 7	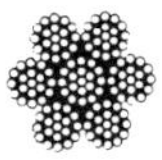 7 x 19	 1 x 19 **	 7 x 7 †	 7 x 19 *	
1/16	500	480		500	480		1/16
3/32	1200	920	1000	1200	920	920	3/32
1/8	2100	1700	2000	2100	1700	1760	1/8
5/32	3300	2600	2800	3300	2400	2400	5/32
3/16	4700	3700	4200	4700	3700	3700	3/16
7/32	6300	4800	5600	6300	4800	5000	7/32
1/4	8200	6100	7000	8200	6100	6400	1/4
9/32	10300	7600	8000	10300	7600	7800	9/32
5/16	12500	9200	9800	12500	9000	9000	5/16
3/8	17500	13100	14400	17500	12000	12000	3/8
7/16	23400	....	17600	23400	....	16300	7/16
1/2	30300	....	22800	29700	....	22800	1/2
9/16	....	....	28500	....	....	28500	9/16
5/8	....	....	35000	....	....	35000	5/8
3/4	....	....	49600	....	....	49600	3/4
7/8	....	....	66500	....	....	66500	7/8
1	....	....	85400	....	....	85400	1

†7/16" and larger made 6 x 7 with I.W.R.C.
*7/16" and larger made 6 x 19 with I.W.R.C.

**Strength efficiency references in this catalog are based on the use of wire rope, not strand. For efficiences of swaged terminals when used with strand constructions (1x19) refer to Sales Department, Wilkes-Barre.

INDUSTRIAL STANDARDS
WIRE ROPE ASSEMBLIES

MATERIAL:

Corrosion Resisting Stainless Steel.

STRENGTH:

These terminals will develop the rated breaking strength of the wire rope.

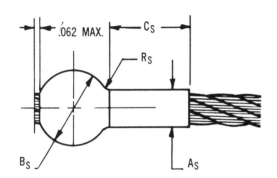

MS-20664 (see first column for Dash Part Number)

DIMENSIONS AFTER SWAGING

MS DASH PART NO. COR. RES. STEEL	NOM. CABLE DIA.	MIN. ALLOW. BREAK. STR. (POUNDS) AIRC. CABLE	As		Bs		Cs REF.	Rs MAX. RAD.
			DIA.	TOL.	SPH. DIA.	TOL.		
C2	1/16	480	.112	+.000 −.003	.190	+.000 −.003	.156	.014
C3	3/32	920	.143	+.000 −.003	.253	+.000 −.003	.234	.019
C4	1/8	2000	.190	+.000 −.003	.315	+.000 −.003	.313	.023
C5	5/32	2800	.222	+.000 −.004	.379	+.000 −.004	.391	.028
C6	3/16	4200	.255	+.000 −.005	.442	+.000 −.005	.469	.033
C7	7/32	5600	.302	+.000 −.005	.505	+.000 −.005	.547	.038
C8	1/4	7000	.348	+.000 −.005	.567	+.000 −.005	.625	.042
C9	9/32	8000	.382	+.000 −.007	.632	+.000 −.007	.750	.046
C10	5/16	9800	.413	+.000 −.007	.694	+.000 −.007	.813	.046

INTERCHANGEABILITY RELATIONSHIP: MS-20663 & AN-663 PARTS IDENTIFIED BY THE SAME DASH NUMBER ARE UNIVERSALLY, FUNCTIONALLY, & DIMENSIONALLY INTERCHANGEABLE.

ALLOWABLE VARIATION: ±.005 ON ALL DIMENSIONS UNLESS OTHERWISE SPECIFIED   GLES ±3°

MFG. IN ACCORDANCE WITH MIL-T-781
CABLE IN ACCORDANCE WITH MIL-W-1511 & MIL-C-5424

DICK RIGNEY Hiring Engineers and Draftsmen

Dick has worked out a little dimensioning test he claims is useful not only for selecting draftsmen but for appraising ability in design engineers as well.

The examination shown as Exhibit 1 — Dick Rigney was used as part of the screening procedure for hiring engineers and draftsmen at Task Corporation of Anaheim, California. In addition to this examination, an applicant was judged on the basis of data given on an application form and personal interviews with Task engineers. Task management did not know of any other company that used such an examination in screening applicants, but thought the test was a most useful part of the overall procedure, which they described as "good though not perfect. We've had only one instance of having a new employee work poorly on the job after scoring well on the test."

The Company

Task was a relatively small company with 140 employees. It manufactured a wide range of products, including electric motors, pumps, fans, blowers, refrigeration systems, flexure joints, and wind-tunnel balances.* One example of a Task motor design appears in the Jack Wireman case (Part 1) in Chapter 2. Electric motors, ranging in horsepower from fractions to hundreds, accounted for the largest share of Task's sales. A special capability of the company was making motors very high in power relative to weight and size. Many of the motors were for aircraft applications, such as running pumps in aircraft and missiles, where extremely high reliability is absolutely essential. To assure that close tolerances were

*A wind-tunnel balance is a device for simultaneously supporting a model to be tested and measuring forces acting on the model. The balance has to be structurally strong so the model will not be torn loose by wind and yet sensitive to the forces. Strain gages in the balance measure forces about various axes, and have to be placed with great skill so the torques and forces can be separately resolved in magnitude and direction.

held with high reliability, the quality control department of the company performed 100 percent inspection on missile-motor parts.

Products of Task were generally quite expensive. Many were made for special applications where small quantities were required. Consequently, production runs were often short, requiring that costs of engineering, tooling, setups, and learning be recovered on only a few units. Engineering costs were frequently high because many special applications imposed extremely difficult constraints. One job, for instance, required design of a 240 hp, air-cooled, continuous-duty electric motor weighing less than 190 lb. The careful machining needed to hold close tolerances, and careful quality control necessary to insure reliability were also expensive. Assembly usually required a high proportion of hand labor, particularly on balances, each of which had to be individually assembled and calibrated by a skilled technician. Materials often were bought at premium prices to assure top quality. Thus a combination of factors pushed upward the cost of Task products. One of the company's balances weighing only ¼ lb sold for $15,000.

The high cost of each unit and the limited quantities of production meant that each product had to be carefully engineered. It was judged too expensive to make numbers of prototypes to shake out the bugs. Furthermore, there generally was not time for much shaking out, since most of the motors were used on special projects running on tight time schedules. There frequently was not time to send motors back to the drawing boards, either. Drawings had to be as correct as possible the first time.

Draftsmen worked with Task engineers in preparing designs. Responsibility for correctness of drawings, however, was primarily with the engineers. Elmer Ward, president of Task and effectively the company's chief engineer, commented on engineering drawing responsibilities: "We encounter engineers who want to stay away from the drafting board, who think they should just be able to make rough sketches and turn the rest over to somebody else. But we can't get good work that way, because this approach forces the majority of detail decisions down to less-qualified and less-responsible individuals. We have to require engineers here to make layout drawings that include all critical details and tolerances. The draftsman then makes more formal manufacturing drawings. But the engineer must go over these drawings and approve them before they are considered finished."

Engineers' responsibility for follow-through at Task went beyond completion of drawings. Unlike some larger companies, where separate specialists write proposals, perform analytical work, design parts, make layouts, and shepherd prototypes through the shop and testing, each Task engineer typically worked his project through all these phases.

Engineers at Task were not especially high-paid, and hours were often long, but the turnover was low. Typically, many of the engineers did not leave promptly at quitting time, and occasionally they worked very late preparing proposals or solving tough problems. One of them commented, "I could work shorter hours and make more money at one of the big companies, but I don't think I'd enjoy it or learn as much. Here I'm not confined to a narrow specialty. I not only work on technical problems, but I get in on sales and also work with the men in the shop as well. The main thing, though, is that here I can carry a project all the way through from concept to working hardware, and that to me is satisfying."

Randy Winters, Task's production manager, said of the company's organization, "We've tried not to organize more than we absolutely had to. We avoided making any organizational charts or job descriptions, hopefully to prevent the occurrence of what every organization seems to strive for, that is, running down hill toward internal stability. Both our workload and our products are subject to so much variation that we can't afford to have people whose jobs are so well-defined that they just do certain things in the way of work, and then figure they're all finished. So much of our work is on a custom-order basis that we have to be able to move people around. Some of our people, who would normally be on salaries in most companies, are on hourly wages here. If they were on salaries they would be less willing to come in to work on Saturdays when the workload is up.

"Our way of operating is frustrating to many people when they first join us. We give them freedom they haven't been used to elsewhere and they feel less secure. When we move them around they get nervous. But if they can hang on for a while, they find out how much they can get done, and then they begin to relax.

"In many engineering problems there is no orderly approach that will work, and it is impossible to plan so that everyone can just work along smoothly. The only time we get superior results is when we have

to have them; and this usually happens at the last minute. This process may be uncomfortable, but we think it is relatively efficient because we generally get the answer we want the first time around, and we have solved a lot of problems that the big companies couldn't handle. Recently we beat a big company that wanted 6 months and $50,000 for a certain type of motor. We built it in 1 month for $23,000. This big company was probably just like several of them that have the pressure so low they are virtually paralyzed. They have huge rooms full of engineers who can't get anything done. We sometimes wonder how they survive."

Mr. Ward shared Mr. Winters' enthusiasm for Task's engineering-design abilities: "About three years ago we started making a tiny motor for a company that will use it in running a calculator. The calculator required a motor that was very small, but quite high-powered. The calculator company had one prominent motor company make them a motor, but the most powerful one they were able to produce within the required size was not powerful enough to run the machine. After seeking help from all the motor companies they knew, and finding that none of them could do the job, the calculator company sent a scatter letter to all the companies they could identify who might be able to build it. We happened to get one of their letters, undertook to solve the problem, and did. Our motor was within the required size limits and gave twice the required power. The customer has since been trying to find another company able to make an equivalent motor as a second source of supply, but without success so far."

The Dimensioning Test

The importance of correctness in engineering drawings prompted Richard Rigney, a Task engineer, to prepare the dimensioning test shown as Exhibit 1 — Dick Rigney. He noted that some of the most annoying problems of prototype machines that did not function properly were caused by improper fits of finely finished and expensive parts. He often traced the source of the difficulty to dimensioning of fits and clearances incorrectly stated on blueprints. He proposed that draftsmen and engineers be given a proficiency test on dimensioning. In the test he included the sorts of problems that seemed to give the most difficulty.

No attempt was made to include tricks or hidden traps in the problems. Mr. Rigney believed, however, that the test indicated more than merely ability at dimensioning, classifying it as "really sort of a simple intelligence test."

Mr. Ward also saw it as more than simply a dimensioning test, saying that its broader function was to test a man's attentiveness to details in general.

"Attention to details is something more engineers need to learn," he said. "It's not enough to do a good job of 'big-picture' thinking on designs. We always get that right. Our real bugaboos are the small things we overlook. People can easily be fooled by paper designs, but you can't fool hardware. It reacts exactly according to the design, details included."

The dimensioning test was given to an applicant with the injunction that he read the instruction sheet *carefully,* and do exactly what it said and nothing else. Both draftsmen and engineers were given the test, and typically the scores of one group were about the same as those of the other. A normally acceptable performance, according to Mr. Rigney, was to make about two or three mistakes. There was no limit set on time, but it was observed that those men the company hired had usually finished in 15 min or less.

Discussion Questions

1 Complete the dimensioning test developed by Dick Rigney (Exhibit 1 — Dick Rigney). How long did it take you? What do you think it measures?

2 What do you think of Task's procedure in selecting engineering applicants? Explain.

3 Which of the other engineers you have read about here do you think would do best on Dick's test, and which one poorest? Which do you think would contribute most to the company, and which least? On what does your decision depend?

4 What do you see as the pros and cons of the engineering management philosophy at this company?

5 Evaluate Warren Deutsch, whose résumé appears in the Warren Deutsch case (Part 1) in Chapter 7, as an applicant for a job similar to that of Jack Wireman at Task Corporation.

Exhibit 1 — Dick Rigney

TASK CORPORATION

NAME_____DATE_____

This test is intended to indicate proficiency in arranging dimension to *meet specific design goals with minimum tolerance buildup.* The objective is to dimension all the axial lengths required to meet the design goals A thru I shown on Figure 1. The nominal dimensions are given by station lines. A standard tolerance of $\pm.010$ must be used with the exception of those required to maintain goal "A" (the shaft end play held within .010–.020). The *minimum number* of close tolerance dimensions must be used to maintain goal "A".

All design goals are listed in their order of decreasing importance. The requirements of "A" should take precedence over "B"; "C" over "D"; etc.

Arrange all dimensions for minimum tolerance buildup, i.e., minimum number of dimensions to locate the required surface.

Priority requirements:

1. "A": shaft end play must be held within limits shown on drawing.
2. "B" thru "H": held to minimum buildup.
3. "I": basic dimension as shown by layout should be modified on one or both parts to insure that one surface of the two shown will always clear the other.

Reread the above directions carefully. This examination is not intended to show proficiency in conventional dimensioning practice. The objective is to demonstrate the ability to understand and interpret directions, and to act in strict accordance with these directions. Thoroughness and attention to all details is an absolute requirement.

IT IS IMPORTANT TO FOLLOW THE INSTRUCTIONS CAREFULLY

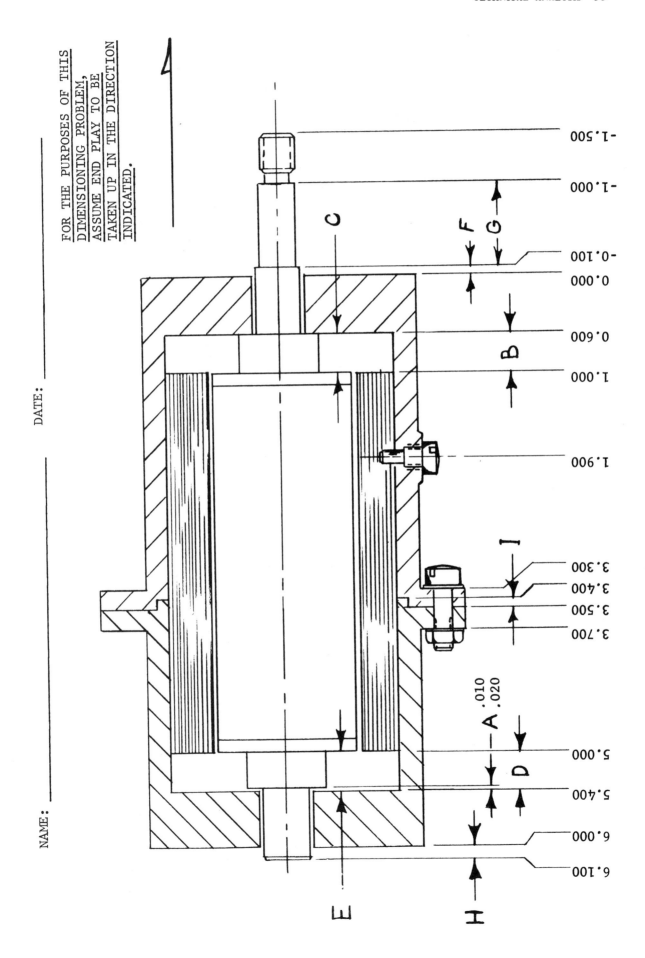

NAME:

DATE:

FOR THE PURPOSES OF THIS
DIMENSIONING PROBLEM,
ASSUME END PLAY TO BE
TAKEN UP IN THE DIRECTION
INDICATED.

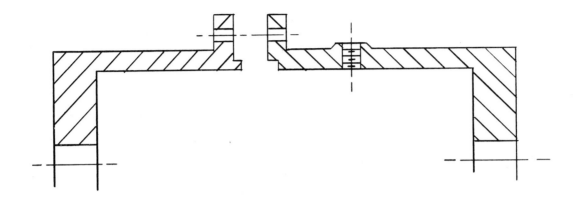

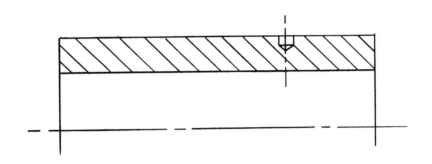

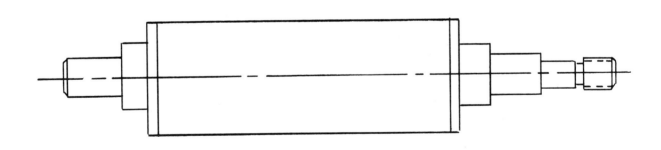

ED FISH   Pushbutton Wiring Circuit

The logic circuit Ed needs to control a tracking antenna doesn't require high-powered math. But thought is required, and some judgment.

"Figuring out how to turn something on and off can be a tough problem," commented Edward Fish, engineer at Philco-Ford's Western Development Laboratories (WDL). In 1963, WDL accepted a contract to design and assemble a 60-ft satellite tracking antenna for the military. Mr. Fish said he was one of about 100 engineers in the department who was to design the antenna. His particular section of the department was responsible for the design of the control system. For example, the section was to develop a combination of devices — such as motors and gears — that would keep the antenna directed automatically toward a moving satellite. The WDL Programs Office, which acted as a liaison between the company and its customer, allowed Mr. Fish's section 3 months to complete its designs. A photograph illustrating this type of antenna appears as Exhibit 1 — Ed Fish.

The Philco-Ford Corporation, which originally established its reputation as a radio manufacturer, began to work with telemetry and control problems at WDL in the early 1950s. After Philco became a part of the Ford Corporation in 1961, WDL grew rapidly, and by 1967 employed about 2,500 people. Much of WDL's work was done under state- and federal-government contracts, and was concerned with receiving and transmitting nonvocal information (telemetry). WDL designed, built, and maintained satellite tracking stations throughout the world. They also designed a system for automatically controlling water flow and levels in a vast complex of dams, canals, and aqueducts to supply two new northern California power plants. More recently they had contracted with the Bay Area Rapid Transit District to design and install a communication system to control the movement of trains in a storage, maintenance, and repair facility.

The satellite tracking antenna on which Mr. Fish was working was to operate in four different modes. In the "track" mode, the antenna would automat-ically follow a moving satellite (see Exhibit 2 — Ed Fish). In the "slave" mode, the antenna's position would be controlled by a computer containing information about the satellite's predicted location. In the "search" mode, the computer would be programmed to cause the antenna to scan the sky in a systematic fashion. In the "manual" mode, the antenna would be free to turn so that operators at the antenna site could rotate it to a convenient position for maintenance work.

The antenna's operation was to be controlled by four push buttons mounted on a panel in the tracking station control room. In describing the operation, Mr. Fish said, "When power is first turned on, operation is to be automatically in the 'manual' mode so that the antenna will not move until a button is pushed. When a mode is started by pushing a particular button, the previous mode is to be disconnected. Pushing the 'track' button, for example, must disconnect all other modes and simultaneously energize a relay (which requires 2 amps) at the antenna site to initiate the 'track' operation. Once initiated, a mode is to continue until another button is pushed. Also, it has to be possible to select the modes in any order."

At the beginning of the tracking antenna project, Mr. Fish spent about a week writing logic equations* to clarify some of the numerous military specifications for the antenna system. He would work out an equation, then explain to the Military Programs Office what it implied about the operation of the antenna. Whenever the response was "that's not what we really meant," he would modify his equation. "If specifications are too precise," he commented, "there's no room for product improvement; but if they're too vague, customers don't always get what they want."

During the week in which he worked on the logic equations, Mr. Fish spent part of his time selecting

---

* An example of a logic statement in such an equation might be; "If switch A and switch B are closed, then switch C is to be open." Logic equations use symbolic notation to express such statements.

push-button units. "If we had needed many sets of buttons, it would have been cheaper to use mechanically operating units rather than electrically operating ones," Mr. Fish said. "Since we don't manufacture most parts used in our products, we would have had another company make the buttons for us. It might have taken quite a bit of time to write specifications for that company." Mr. Fish knew that all parts purchased for a military project were subject to military approval before they could be ordered. Also, the Philco-Ford Quality Assurance Program required a reliability inspection of all purchased parts before approving them for use by Philco-Ford engineers. He expected that these circumstances might stretch out by several months the time needed to obtain the push buttons.

From his previous experience in buying parts, Mr. Fish remembered that several companies kept push buttons with electrically operated holding coils* in stock. He guessed delivery time would be about 3 weeks. He recalled that his company had previously used some of these units and found them satisfactory. He decided to send for catalogs describing the buttons to see if there were any important differences. Inspecting the catalogs, he discovered that units manufactured by Kontroll Industries and those made by Quikswitch Company† were suitable and quite similar to each other. Both units were priced at around $20.

He also considered several human factors. For example, the military specifications required that the buttons must light up when pushed down, and each button had to contain at least two bulbs so that if one bulb burned out, the button would still be lighted. The military also required taking precautions to prevent personnel from contacting potentials greater than 30 V. According to the catalog, both button units operated on less than 30 V.

Philco-Ford's Human Factors group, whose job it was to inspect all products to make sure they would be easy for people to use, judged the button units acceptable. Both Kontroll and Quikswitch provided removable caps in various colors to cover the bulbs. They also provided translucent screens that could be imprinted with appropriate legends, such as "track" and "slave." Mr. Fish noted that the legends were easy to read even when the buttons were not lighted. He decided to use green caps on the button lamps because green normally signifies an "on" condition. However, he decided to order a variety of colored caps just in case the Human Factors group would not approve of his color choice.

* The holding coils hold the switch on after the button is pushed. When the switch is closed, current flows through the coil to produce a magnetic force. Pushing a different button then should interrupt the current to the holding coil and allow the switch to open again.
† Corporate names are pseudonyms.

Upon further comparison of the catalogs from Kontroll and Quikswitch, he concluded that it was more difficult to change the light bulb in the Kontroll unit because a special tool was necessary. The Quikswitch unit, on the other hand, could be disassembled by hand. The light bulbs could then be removed from the clips holding them. Furthermore, Quikswitch had a warehouse in Los Angeles, whereas Kontroll Industries was located in Chicago. He therefore expected that he could get faster delivery from Quikswitch. Taking all these factors into consideration, he decided to use the Quikswitch unit.

According to the catalogs, button units came in sizes that depended upon their operating voltages. Quikswitch units operated from 6 V, 12 V, or 28 V. Kontroll's switches operated from 6 V, 28 V, or 48 V. The higher voltage units were more expensive than the lower ones. The bulbs operated from 28 V. Mr. Fish wanted to use a voltage that would be compatible with Kontroll's units in case something should happen to the Quikswitch buttons and he could not obtain replacements from them. He noted that strikes or high demand, which could usurp a company's supply of stocked items, might create such a situation.

Heat dissipation was not expected to be a problem since the buttons would be surrounded by a lot of air when mounted in the control panel and would not be close to any other equipment. Mr. Fish decided to use button units that operated from 28 V. He told his coworkers that he needed a 24-V power supply. The lower voltage would give the lamps a longer life, and he thought the button relays would still operate at 24 V. Relay contact deterioration would, he expected, be lower at reduced voltage.

Next Mr. Fish considered how to connect the switches in the button units so they would operate according to the military specifications. The Quikswitch catalog showed that the button units were equipped with from one to four single-pole, double-throw switches. He hoped to use only two switches in each unit because this would simplify the wiring. He wanted to make his connections as symmetrical as possible; that is, he wanted to make the connections between the first and second buttons similar to the connections between the second and third buttons, and so forth, as closely as possible. He had seen other engineers design button connections this way and knew that a symmetrical design would make it easier to add more buttons later, if they should be needed. He thought such a design might also help prevent wiring mistakes.

"Although the logic equations clarified the specifications, I had to use trial and error to determine the proper switch connections," said Mr. Fish. He regarded his problem as one of "pure and simple logic." ["It's not what you learn in school, maybe, but it's important to be able to do this sort of thing," he commented.] "In this case what you're turning on and off isn't important. It could be anything. You sit down with this kind of logic problem and you just fiddle. You fiddle until you come up with something that will work. So far, I have the 'track' and 'manual' buttons working properly, and extending the design to the other modes should be straightforward." (See Exhibit 3 — Ed Fish.)

"After I've finished the circuit design, I'll have to make sure the contact rating on the switches will not be exceeded," Mr. Fish added. "This rating tells the maximum amount of current that can pass through a contact without damaging it. The manufacturer's catalog lists this rating as 5 amp and says the solenoid in each unit uses 1.7 watts at 28 V dc. The catalog also says that tubular, single-contact, midget-flanged lamps with bulb diameter of $7/32$ in. must be used in the button units."

Mr. Fish planned to check another catalog to determine how much current such lamps draw at 28 V dc.

Discussion Questions

1 Carry forward the work on Ed Fish's project as far as you can within the limitations of your time and the case information.
2 What possible problems should be anticipated in designing the switch circuit and what should he do about them?
3 Comment on Ed's procedure on the project thus far.

Exhibit 1 — Ed Fish

Exhibit 2 — Ed Fish   Modes of antenna operation

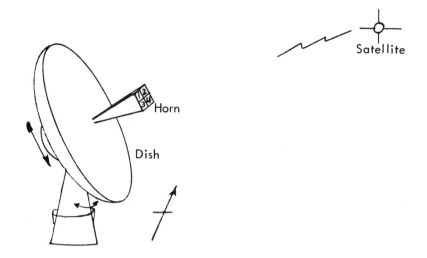

The antenna or "dish" can tilt in a vertical plane or rotate in a horizontal plane. The horn (Fig. 2a) is divided into four sections that separately receive signals from the satellite. If the total signal received by sections 1 and 3 is greater than that received by sections 2 and 4, then the antenna is to be rotated toward the south until there is no difference in the signal power received. Likewise, if the signal power received by sections 1 and 2 is greater than that received by sections 3 and 4, then the dish is to be tilted upward until the difference in signal power is zero.

A device can be designed that will compare the signals received and generate a voltage that is directly proportional to the difference in signal received. (Fig. 2b.) This voltage, in turn, can be used to drive a motor. The motor will produce a torque proportional to the applied voltage. This torque can then be used to correct the position of the dish. Errors in positioning are "fed back" to the input; hence the operation is "automatic" and the antenna will track the satellite across the sky.

When the *track* button is depressed, some of the current through that button unit flows through a solenoid located at some distance from the control panel itself. Current in this remote solenoid then operates another relay that connects the voltage signals from the horn to the system of motors, gears, and amplifiers that controls the system.

Likewise, depressing the *slave* and *search* buttons causes the operation of corresponding distant relays that connect a computer to the control system. The *manual* button does not operate a distant relay since the control system is not automatically supplied with power in the manual mode.

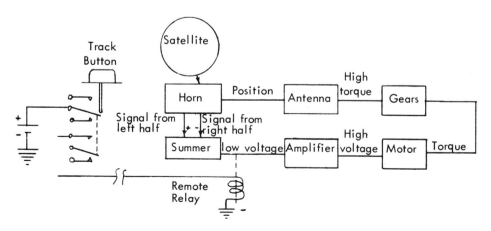

Exhibit 3 — Ed Fish

REQUIREMENTS

Condition	Lights	Mode
Power off	All off	
Power turned on	All off	MANUAL
Button A pushed	Light A on; all others off	Mode A
Button B pushed	Light B on; all others off	Mode B

Note: Only one mode may be on at a time. Modes may be selected in any order.

BUTTON OPERATION

The switching arms of these switches are normally in contact with points "A". When a button is pushed down, both its swinging arms are mechanically forced away from points "A" and against points "B" by the rod (dotted line). If current is flowing in the solenoid, the arms will be magnetically held against the "B" points until current ceases to flow. Then they will return to the "A" positions. The magnetic force caused by current flowing in a particular solenoid is not great enough to pull a button down, however. The force merely holds the button down after it has been manually depressed.

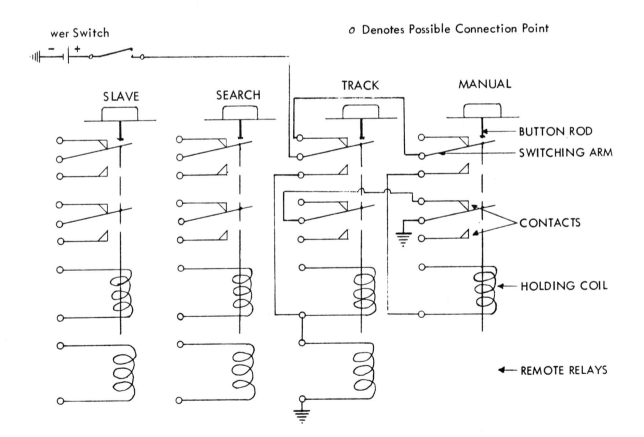

General Electric Miniature, Subminiature Lamps

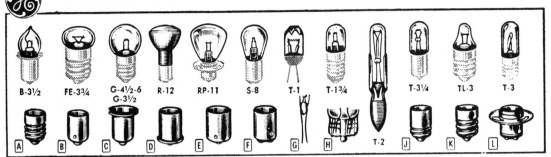

B-3½ FE-3¾ G-4½-6 / G-3½ R-12 RP-11 S-8 T-1 T-1¾ T-3¼ TL-3 T-3
A B C D E F G H T-2 J K L

ABBREVIATIONS USED TO DESCRIBE BASE AND BULB STYLES

BASES: D.C. Bay., double-contact bayonet. D.C. Pref., double-contact, prefocused. Mid. Screw, midget screw. Min. Bay., miniature bayonet. S.C. Bay., single-contact bayonet. S.C., F., single-contact, flanged. S.C. Mid., F., single-contact midget, flanged. Tel. Slide, telephone slide. BULBS: B, lemon. G, globe. FE, flat end. R, reflector. S, straight end. T, tubular. TL, lens end, tubular.

GENERAL-PURPOSE MINIATURE LAMPS

Numbers in Bulb Style indicate bulb diameter in eighths of an inch. Example: S-8 is 8/8ths, or 1" in diameter. *Switchboard slide-type base. †Similar to bulb TL-3; #Similar to bulb G-4½. Wt., 2 oz.

Stock No.	Mfr's Type	Volts	Amps	Fig.	Base Style	Bulb Style	EA.	PKG. OF 10
60 A 7300	PR-2	2.4	.50	C	S.C., F	B-3½	.19	1.16
60 A 7301	PR-3	3.6	.50	C	S.C., F.	B-3½	.19	1.16
60 A 7302	PR-4	2.3	.27	C	S.C., F.	B-3½	.19	1.16
60 A 7303	PR-6	2.47	.30	C	S.C., F.	B-3½	.19	1.16
60 A 7304	PR-7	3.7	.30	C	S.C., F.	B-3½	.24	1.44
60 A 7305	PR-12	5.95	.50	C	S.C., F.	B-3½	.19	1.16
60 A 7306	PR-13	4.75	.50	C	S.C., F.	B-3½	.19	1.16
60 A 7367	6	6.4	3.0	F	D.C. Bay.	S-8	.34	1.99
60 A 7307	12	6.3	.15	...	Min. 2-pin	G-3½	.24	1.42
60 A 7308	13	3.7	.30	A	Screw	G-3½	.16	.99
60 A 7309	14	2.5	.30	A	Screw	G-3½	.16	.99
60 A 7310	24E	24	.035	*	Tel. Slide*	T-2	.63	3.74
60 A 7369	24X	24	.035	*	Tel. Slide*	T-2	.63	3.74
60 A 7312	39	6.3	.36	E	S.C. Bay.	T-3¼	.34	1.89
60 A 7313	40	6.3	.15	A	Screw	T-3¼	.14	.81
60 A 7314	41	2.5	.50	A	Screw	T-3¼	.21	1.27
60 A 7315	43	2.5	.50	E	S.C. Bay.	T-3¼	.21	1.27
60 A 7366	44	6-8	.25	E	S.C. Bay.	T-3¼	.14	.81
60 A 7316	45	3.2	.35	E	S.C. Bay.	T-3¼	.17	1.06
60 A 7317	46	6.3	.25	A	Screw	T-3¼	.14	.81
60 A 7318	47	6.3	.15	E	S.C. Bay.	T-3¼	.14	.81
60 A 7319	48	2.0	.06	A	Screw	T-3¼	.21	1.27
60 A 7320	48C	48	.035	*	Tel. Slide*	T-2	.68	4.04
60 A 7321	49	2.0	.06	E	S.C. Bay.	T-3¼	.17	1.06
60 A 7322	50	6-8	1 c.p.	A	Screw	G-3½	.15	.87
60 A 7323	51	6-8	1 c.p.	E	S.C. Bay.	G-3½	.13	.79
60 A 7370	52	14.4	.10	A	Screw	G-3½	.26	1.58
60 A 7324	53	14.4	.12	E	S.C. Bay.	G-3½	.13	.79
60 A 7325	55	6-8	2 c.p.	E	S.C. Bay.	G-4¼	.13	.79
60 A 7326	57	14	2 c.p.	E	S.C. Bay.	G-4½	.13	.79
60 A 7328	67	13.5	4 c.p.	E	S.C. Bay.	G-6	.17	1.01
60 A 7329	67K	13.5	4 c.p.	K	Candelabra	G-6	.43	2.51
60 A 7330	81	6-8	6 c.p.	E	S.C. Bay.	G-6	.21	1.26
60 A 7331	82	6-8	6 c.p.	F	D.C. Bay.	G-6	.21	1.26
60 A 9707	93	12.8	1.04	E	S.C. Bay.	S-8	.30	1.79
60 A 7332	112	1.2	.22	A	Screw	TL-3	.16	.99
60 A 6737	137	6.3	.25	E	S.C. Bay.	G-3½	.17	1.06
60 A 6738	157	5.8	1.10	A	Screw	G-6	1.00	5.95
60 A 7334	158	12	.24	H	Wedge	T-3¼	.20	1.14
60 A 7335	159	6.3	.15	H	Wedge	T-3¼	.21	1.26
60 A 7336	222	2.2	.25	A	Screw	TL-3	.16	.99
60 A 7337	223	2.2	.25	A	Screw	FE-3¼	.19	1.16
60 A 7338	224	2.15	.22	J	Special	†TL-2¾	.19	1.16
60 A 7339	233	2.3	.27	A	Screw	G-3½	.19	1.16
60 A 6739	240	6.3	.36	E	S.C. Bay.	G-3½	.21	1.26
60 A 7340	248	2.5	.80	A	Screw	#G-5½	.26	1.58
60 A 7373	259	6.3	.25	H	Wedge	T-3¼	.19	1.16
60 A 7341	313	28	.17	E	S.C. Bay.	T-3¼	.30	1.78
60 A 6741	324	3	.19	G	Wire	T-1¼	.58	3.45
60 A 7342	327	28	.04	D	S.C.Mid.F.	T-1¾	.70	4.15
60 A 7343	328	6	.20	D	S.C.Mid.F.	T-1¾	.54	3.25
60 A 7374	330	14	.08	D	S.C.Mid.F.	T-1¾	.78	4.61
60 A 7145	331	1.35	.06	D	S.C.Mid.F.	T-1¾	.84	5.01
60 A 7146	334	28	.04	...	Mid.Groove	T-1¾	.76	3.98
60 A 6742	337	6	.20	...	Mid.Groove	T-1¾	1.55	9.09
60 A 7375	344	10	.014	D	S.C.Mid.F.	T-1¾	1.46	8.46
60 A 7376	345	6	.04	D	S.C.Mid.F.	T-1¾	1.45	8.97
60 A 6743	346	18	.04	...	Mid.Groove	T-1¾	1.26	7.53
60 A 7377	363	14	.20	E	S.C. Bay.	G-3½	.21	1.26
60 A 8900	367X	10	.04	D	S.C.Mid.F.	T-1¾	1.60	9.60
60 A 8901	368	2.5	.20	D	S.C.Mid.F.	T-1¾	.87	5.30

GENERAL-PURPOSE MINIATURE LAMPS (Cont'd)

Stock No.	Mfr's Type	Volts	Amps	Fig.	Base Style	Bulb Style	EA.	PKG. OF 10
60 A 6744	370	18	.04	D	S.C.Mid.F	T-1¾	1.26	7.53
60 A 6745	386	14	.08	...	Mid.Groove	T-1¾	1.26	7.53
60 A 7378	405	6.5	.50	A	Screw	G-4½	.27	1.67
60 A 7379	406	2.6	.30	A	Screw	G-4½	.22	1.33
60 A 7380	407	4.9	.30	A	Screw	G-4½	.20	1.22
60 A 7344	425	5	.50	A	Screw	G-4½	.19	1.16
60 A 7346	432	18.0	.25	A	Screw	G-4½	.21	1.26
60 A 7347	433	18	.25	E	S.C. Bay.	G-4½	.26	1.58
60 A 7382	502	5.1	.15	A	Screw	G-4½	.16	.99
60 A 7383	509K	24	.18	K	Candelabra	G-6	.26	1.58
60 A 6600	1133	6	3.91	E	S.C. Bay.	RP-11	.33	2.29
60 A 7384	1383	12	20w	E	S.C. Bay.	R-12	2.65	15.81
60 A 7349	1445	18	.15	E	S.C. Bay.	G-3½	.19	1.13
60 A 7350	1446	12	.20	A	Screw	G-3½	.24	1.45
60 A 7351	1447	18.0	.15	A	Screw	G-3½	.19	1.16
60 A 7353	1458	20	.25	E	S.C. Bay.	G-5	.26	1.58
60 A 7385	1474	14	.17	A	Screw	T-3	.34	2.03
60 A 7354	1477	24	.17	A	Screw	T-3	.52	3.13
60 A 7355	1487	12-16	.20	A	Screw	T-3¼	.24	1.45
60 A 7356	1488	14	.15	E	S.C. Bay.	T-3¼	.26	1.58
60 A 7357	1493	6.5	2.75	F	D.C. Bay.	S-8	1.02	6.08
60 A 7386	1630	6.5	2.75	L	D.C. Pref.	S-8	1.36	8.00
60 A 7359	1768	6	.20	A	Mid.Screw	T-1¾	.90	5.21
60 A 7388	1769	2.5	.20	A	Mid.Screw	T-1¾	.84	5.01
60 A 7360	1813	14.4	.10	E	S.C. Bay.	T-3¼	.24	1.45
60 A 7361	1815	12-16	.20	E	S.C. Bay.	T-3¼	.19	1.16
60 A 7389	1816	13	.33	E	S.C. Bay.	T-3¼	.23	1.38
60 A 7362	1819	28	.04	E	S.C. Bay.	T-3¼	.35	2.07
60 A 7363	1820	28	.10	E	S.C. Bay.	T-3¼	.31	1.84
60 A 7390	1822	36	.10	E	S.C. Bay.	T-3¼	.58	3.48
60 A 7391	1828	37.5	.05	E	S.C. Bay.	T-3¼	.63	3.77
60 A 7364	1829	28	.07	E	S.C. Bay.	T-3¼	.34	2.03
60 A 7392	1835	55	.05	E	S.C. Bay.	T-3¼	.63	3.77
60 A 7365	1847	6.3	.15	E	S.C. Bay.	T-3¼	.17	1.06
60 A 7596	1869D	10	.014	G	Wire	T-1¾	1.10	6.58

100,000-HOUR SUBMINIATURE LAMPS

Last about 12 years! Use where space is critical, service difficult. Solder-dipped leads on wire types. ¼x⅛" dia. * ±20%. 1 oz.

Stock No.	Mfr's Type	Volts	Amps ±10%	Fig.	Base Style	Bulb Style	EACH	PKG. OF 10
60 A 7393	680	5	.060	G	Wire	T1	3.10	18.62
60 A 7394	682	5	.060	D	Sub.Mid.F	T1	3.88	23.17
60 A 7395	683	5	.060	G	Wire	T1	3.10	18.62
60 A 7396	685	5	.060	D	Sub.Mid.F	T1	3.88	23.17
60 A 7397	715	5	.115	G	Wire	T1	3.10	18.62
60 A 7399	2128	3	*.0125	G	Wire	T1	4.74	28.23

50,000-HOUR MINIATURE LAMPS

Design life is over 5 years—average life is 50,000 hours. Actual life is determined by environmental conditions: vibration, shock, temperature, voltage fluctuations. Numbers in Bulb Style indicate bulb diameter in eighths of an inch. Av. shpg. wt., 1 oz.

Stock No.	Mfr's Type	Volts	Amps	Fig.	Base Style	Bulb Style	EA.	PKG. OF 10
60 A 7400	380	6.3	.04	D	S.C.Mid.F.	T-1¾	1.74	10.61
60 A 7401	381	6.3	.20	D	S.C.Mid.F.	T-1¾	1.10	6.58
60 A 7402	382	14	.08	D	S.C.Mid.F.	T-1¾	1.26	7.53
60 A 7403	755	6.3	.15	B	Min. Bay.	T-3¼	.26	1.58
60 A 7404	756	14	.08	B	Min. Bay.	T-3¼	.34	1.89
60 A 7405	757	28	.08	B	Min. Bay.	T-3¼	.43	2.51
60 A 7406	2180D	6.3	.04	G	Wire	T-1¾	.95	5.64
60 A 7407	2181D	6.3	.20	G	Wire	T-1¾	.52	3.13
60 A 7408	2182D	14	.08	G	Wire	T-1¾	.63	3.77

ART WHITING   To Weigh a Man in Space

How can a man's weight be determined outside the pull of gravity in a space capsule? Art devised some concepts, chose a couple, and built a jury-rig. How adequate are the results?

In late 1964, Art Whiting, a mechanical engineer at the Space Science Laboratory of Colossal Defense Corporation, was testing a device he had developed for measuring the mass of an astronaut in space. It consisted of a chair with two wheels rolling on a central I-beam, plus two outriggers, as depicted in Exhibit 1 — Art Whiting. As the chair was rolled forward it stretched a spring. When released, the chair was pulled backward by the spring, accelerating until it reached the limit of travel. From the elapsed time, Art expected to deduce the acceleration, and from the acceleration and spring force he expected to deduce the "weight" of the chair and its occupant. He had performed a series of tests with men of different weights in the chair and was now ready to compute their weights from the test data to evaluate the effectiveness of his invention. Then he thought he should turn his attention to how the system could be improved in the light of its purpose.

Start of the Project

Art first learned of the desire for a device that could weigh a man in space when he was called in by his supervisor, Cal James, and introduced to Al Lloyd of the Biotechnology Division. Mr. Lloyd explained that he and an electrical engineer had given thought to a system that would consist of a chair suspended from the ceiling on wires, which would oscillate back and forth against springs. Then the mass of the passenger would be computed by the frequency of oscillation; a heavy person causing a slower oscillation and vice versa. He said his division wished to use a working prototype as the basis for a proposal to the government for a contract to produce the device, and a deadline 10 days hence had been set to make the working prototype. Art was assigned by Mr. James to work on the task, with Mr. Lloyd as project engineer.

Art's background had included studies leading to B.S. (1944) and M.S. (1947) degrees in mechanical engineering at a large Midwestern university. He also had a master's degree in industrial management. He had worked 3 yr for a major home-appliance manufacturer on electromechanical controls such as automatic washing-machine timers, and then for 6 yr as chief mechanical engineer in the tube division of an electronics company. "I left when the company decided to change to a project-structured organization rather than staying functionally structured. This meant that most of the mechanical engineers would report to project managers, rather than to me, and my responsibilities were greatly reduced. Besides, in an electronics company, the mechanical engineer is never king on a job. So I was glad to go to work on more mechanical systems here at Colossal." Art had been with Colossal for 5 yr when the project of finding a method for weighing a man in space was initiated.

"I suggested we consider hanging the chair by linkage like a garden swing to prevent sideways motion," Art recalled, "but Lloyd said we should reach for a working model as soon as possible at the absolute minimum cost. Aside from that we had no given design or performance specifications." The two men considered the objectives of minimizing weight and size to be obvious from the constraints of use in a space capsule, and they felt that an accuracy of within 1 or 2 lb was desirable. NASA soon afterward issued a request for a system with an accuracy of ½ lb, but Mr. Whiting thought this an unlikely possibility. "The limiting factor is that the human body is not a rigid body. You can vary the reading of a bathroom scale, for instance, just by waving your arms when you stand on it.

"The first question," he continued, "was how to get something working and obtain test results in 10 days. To build such a machine that fast in your garage is easy. But in a big organization you have the additional problem of fighting red tape. To hang the wires from the ceiling meant first I had to have a safety inspector come and certify that it wouldn't weaken the beams or pull the roof down. I had to arrange it with the maintenance people, who are in charge of the buildings. Union rules prohibit our own lab technicians from working on the building itself. The machinists have one union, the carpenters have one, and the electricians have another. It's simpler for an electronics engineer. He can walk out to a bench, pick up a soldering iron and make any kind of circuit he wants without checking with anyone."

Art's first action was to make a rough sketch of the proposed system. A chair was to be hung from the 19-ft-high ceiling by wires. Coil springs, first stretched 10 in. each and then attached to the front and rear, pulled it back to center when it was displaced. The experimenter was to push it away from center, then release it to allow chair and passenger to oscillate back and forth across center as the springs tried to restore it to the neutral position. To time the period of oscillation, Art obtained a stopwatch. He wrote requests to the company woodshop to build the chair, to the maintenance department to hang the wires, and then visited a local hardware store where he bought some springs. "I didn't know what frequency we should try for," he said, "so I bought two different sizes of springs. As it turned out, the weaker springs damped out too fast, so we couldn't time more than a few oscillations.

"I did some of the construction work myself," he continued. "I had access to some simple tools, including a drill, a grinder, a band saw, and hand tools. So I bought some parts at the hardware store, and made a few parts, like brackets and spring mounts. I figured it was cheaper and it would save time. Besides, in working on the contraption myself I think I got a better feel for the problems of making it work. As it turned out, everything I personally made worked out alright. We had the chair hooked up in just a little over 10 working days ready for the first trial run." (See Exhibit 2 — Art Whiting.)

"We ran it a couple of times to see that it worked and ran calculations with the simple frequency equation:

$$ f = \frac{1}{2\pi} \sqrt{\left(\frac{A}{L}\right)^2 \frac{2Kg}{W}} $$

to see what results would come out. The correlation was somewhat off, as I expected, because the equation ignored certain effects, such as friction. I pulled out one of my college textbooks on vibrations and worked out a more complete equation. It had three terms, one for the spring action, one for the 'pendulum effect' caused by the fact that the chair swung from the ceiling, another for friction. Some of the guys thought there might be a 'hysteresis effect' in the springs, but I didn't think it was significant." The formula then was

$$ f = \frac{1}{2\pi} \sqrt{\underbrace{\frac{g}{L}}_{\text{pendulum}} + \underbrace{\left(\frac{A}{L}\right)^2 \frac{2Kg}{W}}_{\text{spring}} - \underbrace{\frac{C^2g^2}{4W^2}}_{\text{damping}}} $$

where:

g = gravitational constant, 32.2 ft/sec^2
L = distance from point of support to center of gravity (ft)
A = distance from point of support to the springs (ft)
W = weight (lb)
C = damping coefficient (lb-sec/ft)
K = spring constant (lb/ft)

During experimental runs the chair was first displaced not more than 9 in., then released. The stopwatch was used to time 10 oscillations, which typically took about 20 sec. He noticed that when the man in the chair was relaxed, his head and shoulder would swing at each end of the stroke, and the result was a changed period of oscillation. He also found that by sitting off-center the chair could be made to swing from side to side, and again a longer period would result. Consequently, he set a rule that the "passenger" should sit in the center of the chair and as rigidly as he could during all tests. Exhibit 3 — Art Whiting shows a plot of test results that Art described in a memo as follows:

Results of Initial Experiment: Repeatability within 0.4 sec was obtained over 10 cycles. This corresponds to a total variation in apparent subject weight of 3 lb (i.e., ± 1.5 lb). Variation of any experimental point from the curve was within ± 4 lb. This range includes the foregoing error.

Acceleration System

"Lloyd didn't think this was accurate enough, so he asked me to prepare a brief written discussion of it as part of a proposal to higher management that we do more work on it. I thought we should consider more alternative schemes at this point, so I did a morphological analysis. I listed alternative ways of determining weight along one side of a page, and criteria for performance along another side so we could compare them." (A copy of this page appears as Exhibit 4 — Art Whiting, along with some of Art's description of alternatives.)

Art explained that he and Mr. Lloyd were generally in agreement on these criteria, excepting the one concerning "low G-force fields." By these, Art meant the force acting when the capsule was in the gravitational pull of a planet or when it was accelerating. In his view these effects might be significant in some missions. Since no particular capsule or mission had been specified for the weighing device, he thought they should be assumed. Mr. Lloyd, however, thought they were not important and that the criterion concerned with low force fields should be ignored.

"When Lloyd said the low G-force effects should be ignored it made the alternative of measuring by Newton's Second Law of Motion ($F = Ma$) look best," Art observed, "so we decided to try that as a way of getting higher accuracy. We also decided it would be best to eliminate the pendulum effect by simply rolling the chair back and forth on wheels."

At this point Art roughed out a sketch of the proposed system and also made an estimate of the cost in materials and labor. Excerpts from his notebook showing these estimates appear in Exhibit 5 — Art Whiting. "Each of us has to keep track of time spent working on different projects using an authorized charge number," he explained. "To charge time to a given number there must be approval of management in the Biotechnology Section, granted on the basis of our estimates."

The new system was to use a negator spring, rather than coil springs. The unique feature of this spring is that its restoring force is not proportional to extension, as with coil springs, but rather remains constant. Against this spring the chair was to be pulled from rest like a slingshot, then released. Acceleration then would be determined from the time the chair took to travel a given distance. To stop the chair, Art considered several devices, including elastic ("bungee") cords and automobile shock absorbers. He finally decided to stop the chair against corrugated cardboard during the initial experiments.

In choosing the size of the negator spring, Art said he considered the necessity for a limited track length to fit in a space ship. He thought it important to accelerate and decelerate slowly enough to avoid discomforting the passenger. "We ordered several springs," he said, "so we could experiment to see which was most appropriate. We thought we should plan on a force of less than 1 G, and so somewhat arbitrarily picked a range of $\frac{1}{3}$ to $\frac{1}{5}$ G for our trials.

"We planned to install the spring so it would pull the chair forward. A main advantage of this was that then the chair would serve as a brace to hold the passenger's head and back rigid during acceleration. But then we started asking what would happen if anything broke, and we decided to attach the spring to the back of the chair where it couldn't possibly hurt anyone if it let go."

To time the travel during acceleration, Art used a microswitch triggered by a cam along the course of travel. When the chair began to move, the switch was closed by the cam. After traveling 31.5 in. along the cam it was again opened. An oscillator and electronic counter were triggered by the microswitch to determine time. Art described his construction of the timing system. "I thought the way to do it would be to hook an oscillator to the switches and let the counter keep track of the number of pulses it emitted between start and stop. So I checked out an oscillator and a counter and tried to hook them up according to the instructions that come with the counter. When I had some trouble and asked an electronics engineer for help, he told me the counter already had an oscillator built into it. My circuit was generally all right, but the extra oscillator wasn't necessary."

Next, some passengers were tried in the chair to test the system. Pulling just the chair and passenger with a fish scale, Art measured the frictional drag to be 10 oz with a 170-lb man, and about 5 oz when empty. In making this measurement, he observed that the scale readings were very high at the start of the stroke unless he pulled very slowly and evenly. "We took a series of readings," he said, "then averaged the results. I would guess that the accuracy was within plus or minus half an ounce."

In timing the acceleration test runs, Art was somewhat disturbed to notice that the times varied widely depending upon how the man who released the "slingshot" let it go. By pulling his hand off quickly, a considerably lower time resulted. So testing was stopped, and a solenoid-operated, apartment-door latch, bought at a local hardware store, was installed. This improved consistency of the data, according to Art, and resulted in the times shown in Exhibit 6 — Art Whiting.

Now Art felt he should compute the weight from the test data and see how well it compared to the conventionally measured body weights of the passengers. He began his calculations by assuming that the force acting had several components:

$$F_t = F_s - F_B - F_R - F_i$$

where:

F_t = total force available to accelerate cart
F_s = spring force (12.61 lb)
F_B = drag due to ball bearing friction (lb)
F_R = drag due to rolling friction of wheel on track (lb)
F_i = drag due to inertia of all wheels (lb)

From his measured values of friction he thought that 0.62 lb would be a reasonable approximation for $F_B + F_R$.

To determine F_i he reasoned as follows: for a wheel that accelerates

$$T = I\alpha$$

where:

T = torque (ft-lb)
I = rotational moment of inertia (lb-ft/sec^2)
α = angular acceleration (rad/sec^2)

then, $T = \frac{1}{4}F_iR$ where R = radius of the wheel (ft). Also, $I = \frac{1}{2}mR^2$ where m = mass of the wheel (slugs); and $\alpha = a/R$, where a = linear acceleration. Substituting the last three equations in the preceding one:

$$F_iR = 4(\frac{1}{2}mR^2)(a/R)$$

or

$$F_i = 2\frac{1}{2}ma$$

Art weighed the wheels and found them to be 13 lb, so he approximated F_i at $0.2(2s/t^2)$. The total cart weight with no passenger in it he measured to be 58 lb.

He then believed he was ready to apply Newton's Second Law using the test data to compute passenger weights. "They should be correct within a couple of pounds," he said. "If they're off more than 2½ lb I should probably look for ways to make further refinements in the system. If they're off still more, I should probably think through the whole approach again and consider whether some other system should be tried, and if so, which one."

Discussion Questions

1 Carry forward the work on Art's project as far as you can within the limitations of your time and the case information.
2 State the best decisions you can for Art, indicate the underlying assumptions you had to make, and describe a plan for proceeding from there on.
3 Critique Art's procedure thus far on the project.

Exhibit 1 — Art Whiting

Exhibit 2 — Art Whiting

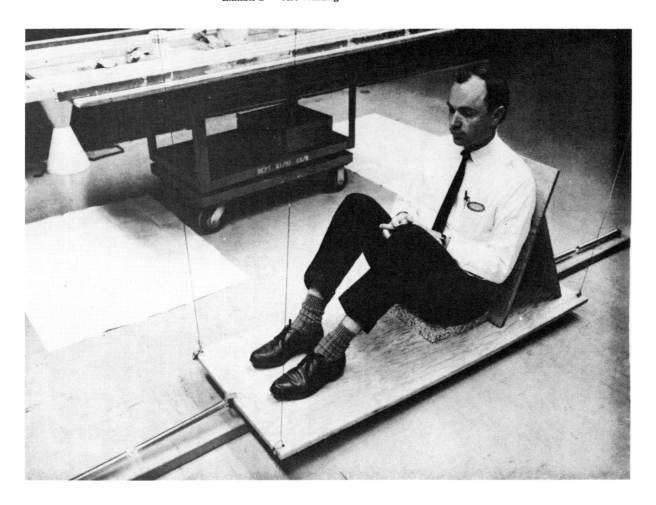

Exhibit 3 — Art Whiting

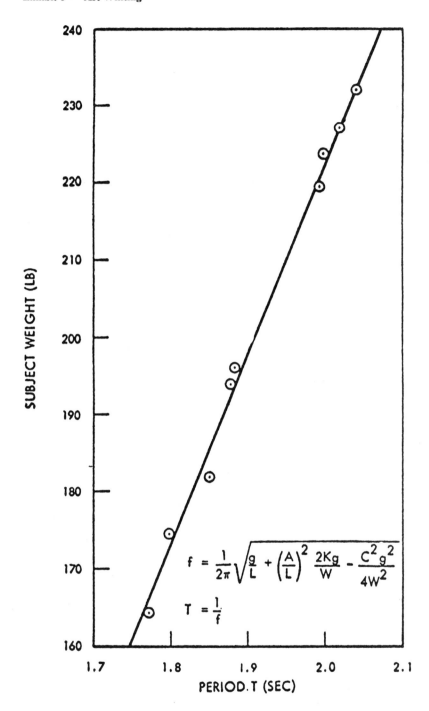

$$f = \frac{1}{2\pi}\sqrt{\frac{g}{L} + \left(\frac{A}{L}\right)^2 \frac{2Kg}{W} - \frac{C^2 g^2}{4W^2}}$$

$$T = \frac{1}{f}$$

Relationship Between Body Mass and Period as Determined by Oscillating Spring-Mass Device (Pendulous Support and Stopwatch Timing)

Exhibit 4 — Art Whiting Art's criteria of comparison for alternate mass measurement systems

MASS MEASURE-MENT SYSTEM	EFFECT OF PRESENCE OF LOW FORCE FIELDS	BEDDING DOWN	APPARATUS SIMPLICITY AND AL-TERNATIVES	SUITABILITY FOR GROUND EVALUATION	EFFECT OF DIFFERENCE IN BEARING FRICTION BETWEEN GROUND AND ORBIT	OPER-ATIONAL CONSIDER-ATIONS
Linear Acceleration (F = Ma)	Disruptive	Little effect	Electronic timer moderately complex; proposed on-board computer could perform function	Ground evaluation satisfactory; may require recalibra-tion for zero-g condition	Could be apprecia-ble; ground calibration chart requires alteration for orbital use	One shot; a 1 man operation
Oscillating Spring-Mass System	No effect unless force field is inter-mittent for brief periods	Greater adverse effect on accuracy than linear ac-celeration	Electronic timer moderately complex; proposed on-board computer could perform function; if electronic timer fails, a stopwatch usable (less accuracy)	Ground evaluation satisfactory; may require recalibra-tion for zero-g condition	Could be apprecia-ble; ground calibration chart requires alteration for orbital use	One shot; a 1 man operation
Inertia-Beam Balance Technique	Effect may be mini-mized by adding pivot	Phased-stroke solution to potential problem	Simplest system	Ground simulation difficult; elaborate means required	None	Trial and error; possible in 2 man operation
Centrifuge	Effect could be compen-sated	No problem	Relatively simple (if centrifuge exists for other purposes)	Ground calibration set-up can simulate zero-g condition	None	Impose scheduling con-straints, unless performed concur-rently with other centrifuge function; involves long operating cycle

ART'S DESCRIPTION OF ALTERNATIVE SYSTEMS

Centrifuge. The equation for centrifugal force is $F = Mw^2R$ where M is mass, w is rotational speed, and R is radius of the path of the mass center. Since R enters this equation linearly, a small error in assumed mass-center location will not be as serious as in the case of the torsional pendulum. However, angular velocity enters quadratically, and will therefore have to be controlled closely. It will not be difficult to measure centrifugal force, but a seat or couch having radial freedom of movement along one arm of the centrifuge will be required.

Impulse Momentum. The equation for this impact action is $F(\Delta T) = M(\Delta V)$. F is a force that is applied over a time period T. M is the mass that is being accelerated from its initial velocity to its final velocity ΔV. In other words, $F(\Delta T)$ represents an impulse which changes momentum $M(\Delta V)$. Both F and T will be difficult to determine because of their transient nature. The presence of random g forces will adversely affect accuracy; therefore, precautions will be required to ensure that no such disruptive forces exist during astronaut mass determination. Further, in a zero-g environment, an astronaut will not be bedded down firmly on the apparatus at the first instant when acceleration starts. Even though he is strapped or clamped in the apparatus, his weightless configuration relative to the apparatus will not be identical to that which his body will assume after inertial forces are present. These considerations make the impulse-momentum method quite impractical.

Conservation of Momentum. The equation for this phenomenon is $MV_1 + mv_1 = MV_2 + mv_2$. It represents a controlled collision in which a smaller known mass (m) traveling with a known initial velocity v_1 strikes the unknown mass M, which is standing still ($V_1 = 0$). A device will be provided to lock the two masses together upon impact so that $V_2 = v_2$. The equation thus becomes $M = (m/v_2)(v_1 - v_2)$. To determine the unknown mass M, the final velocity v_2 must be measured. This can be accomplished by timing the passage of the locked-together masses over a known distance. The disadvantages of this conservation-of-momentum method are that an appreciable auxiliary mass m is needed, and that a considerable jolt will be caused by the impact. Since the presence of random g forces can cause inaccuracies, mass determination should not be attempted unless such forces are absent. The foregoing remarks regarding bedding down apply also to the conservation-of-momentum method and make it impractical.

Linear Acceleration. This method is represented by the equation $F = Ma$ where F is a known, constant force; M is the unknown mass; and a is an acceleration which must be measured. This acceleration can either be measured directly, using an accelerometer, or indirectly by determining the time t to traverse a known distance s; thus, $a = 2s/t^2$. Obviously, time must be measured very accurately since it enters to the second power. If any random g forces are present, they can cause inaccurate results. The bedding down aspect previously noted (the impulse-momentum system) does not apply very strongly to this system ($F = Ma$) because it involves a far longer period of acceleration, and the bedding down will take place at the start of motion. Further, the acceleration will be low, and, consequently, the bedding down effect will be smaller in magnitude.

Frequency of a Spring-Mass System. The frequency of an oscillating spring-mass system is

$$f = \frac{1}{2\pi} \sqrt{\frac{2K}{M}}$$

where

f = frequency (cps)
K = known spring constant
M = unknown mass

One need only provide a single-degree-of-freedom mounting for the unknown mass and a spring system to keep the mass oscillating in simple harmonic motion, and measure the system frequency. Timing can be performed with great accuracy because an average of a large number of cycles can be obtained. Further, the presence of random g forces will have little or no effect on accuracy. With respect to the bedding-down effect, the spring-mass system will probably not be quite as good as the straight linear acceleration system.

Inertia Beam Balance. A parallel platform-type of beam balance (i.e., a commercial weighing scale) can be employed by applying a momentary acceleration in line with the knife edge so that the resulting inertia force acts to replace gravitational attraction in the weighing function. The momentary acceleration can be obtained by moving the balance upward, using a man-powered lever. This system has the great advantage that no additional instruments of any kind are required since there is no need to measure time, force, acceleration, etc. The presence of a random g force will tend to reduce accuracy slightly, but its existence can be detected by making a trial balance without any subject on the platform. In order to overcome the undesirable effects of the bedding-down problem, a rather long accelerating stroke will be required, perhaps with a provision for performing the balance only during the latter part of the stroke.

Exhibit 5 — Art Whiting

Nov. 24, '64

Stripped-down version of F=Ma

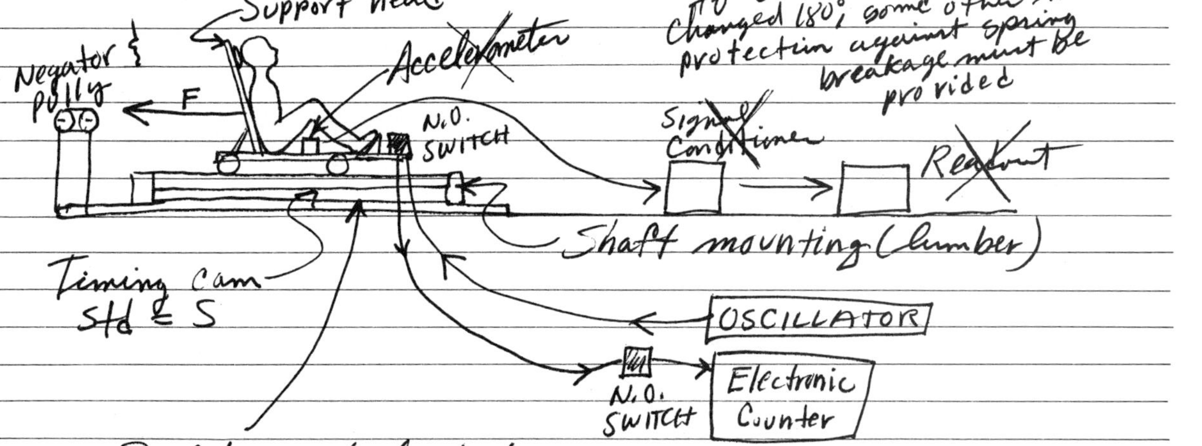

If direction of acceration is changed 180°, some other means of protection against spring breakage must be provided

Support head

Accelerometer

Negator & Pulley

F

N.O. Switch

Signal Conditioner

Readout

Shaft mounting (lumber)

Timing Cam Std & S

OSCILLATOR

N.O. Switch

Electronic Counter

Ball brg. wheels I beam.
(also start latch & stop
Snubber &) latch

EST. Purch Items	$			
		30	30	— Shafting
Negator & Pulley	90	50	50	
Linear Ball brg & Shaft	175	70	76	
Accelerometer	350	—		
Switch	25	25	25	
Signal conditioner Components	200	—	—	
	$840	$175	181	

Largest Permissable Travel ?
3 ft.

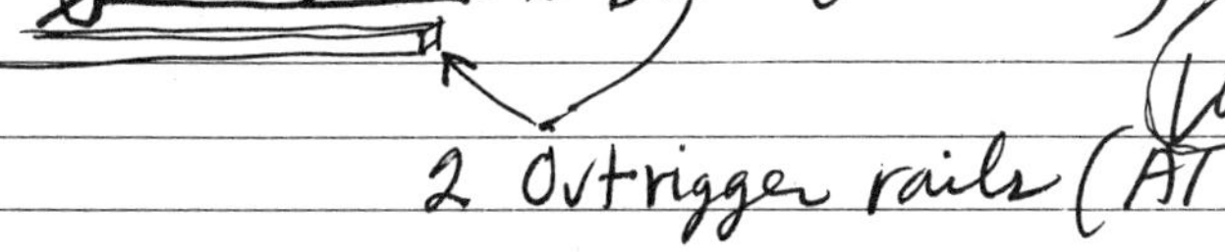

I Beam (w. Shims)

2 Outrigger rails (AT Channels)

Notes: Wt. & pulley may be used as backup
for negator.
TRACK MUST BE LEVELLED.
Negator and pulley
Ball brg. wheels } will have inertia

Nov. 25, '64

Estimate of <u>MAN DAYS</u> to design, build & demonstrate mass determination by use of Newton's Second Law of motion (measuring time instead of acceleration)

	For Combined Equipment	Switch and Precision Timer	Ball Bearing mounting Start Latch & Stop Snubber	Negator spring & Pulley	TOTAL
ENGR — OTHER					
Prelim. design		1	4	2	7
Detail design		1	6	3	10
Order purch. items		2	3	1	6
Drwgs.		1-2	3-6	1-2	5-10
Mfg. special equip.		2	7	1	10
Set up equip for test		1-1	2-2	1	4-3
Perform tests	3-2				3-2
Reduce data	2				2
Modify equip.			1-1	1-1	2-2
Rerun tests	2-1				2-1
Reduce data	2				2
Write Report	4				4

Est. Purch. items

Negator & Pulley	$50
Shafting	30
Ball brgs.	75
Switches	25
TOTAL	$180

TOTALS | 13-3 | 6-5 | 19-16 | 9-4 | 47-28

Exhibit 6 — Art Whiting Linear acceleration test data (copied from Mr. Whiting's notes)

				EMPTY CART WEIGHT 58.0 LBS. TRAVEL 31.5 INCHES					
TEST NUMBER	1	2	3	4	5	6	7	8	9
Passenger Weight (pounds)	162½	173½	144¾	271¾	201¼	169	161½	186¾	132¾
Travel	1764	1816	1686	2144	1906	1818	1763	1861	1633
Time	1745	1795	1688	2165*	1926*	1808	1750	1850	1654*
(milliseconds)	1759	1802	1678	2122	1896	1813	1764	1844	1642
	1745	1810	1681	2116	1898	1797	1750	1839	1629
	1765	1806	1679	2113	1906	1806	1760	1842	1625
Average Time (seconds)	1.7556	1.8058	1.6823	2.1238	1.90150	1.8084	1.7574	1.8472	1.6323

* Crossed Out on Mr. Whiting's Data Sheet

JACK WIREMAN

Failure of a Ball Bearing

Part 3

The consultant Jack called, after failure of the "fix" bearing recommended by the bearing company, has blamed different causes from those the bearing company did, but recommends the same solution, a heavier bearing. Now a second, heavier bearing has just failed.

Further recommendations on the bearing problem were submitted in writing, after galling appeared on the second bearing (204HJB1519), by both the Orbit Company and the bearing expert, Mr. Barish. Both recommended that bearings of still higher capacity be used. A number of possible causes of failure were suggested, but there was some conflict between the opinions of Mr. Barish and those of Orbit about the reasons for failure. Mr. Wireman wondered which of the suggested causes of failure he should consider most likely, and what should be done to cure it. He was especially puzzled about why the change from the first bearing, No. 204SST5 to one of higher capacity, No. 204HJB1519, had ended in failure, since both the bearing company and Mr. Barish advocated higher capacity as the answer.

Opinions of the Orbit Bearing Company

After receiving the second set of failing bearings from the first motor, the Orbit Company returned its analysis in a letter dated December 4, 1963. Orbit repeated its earlier injunction, that a bearing of heavier capacity was needed. The failure, Orbit explained, was due to the poor lubricity (their opinion) of Skydrol. As a cure, Orbit recommended that custom bearings, costing from $60 to $175 each in small quantities, be used. A copy of the Orbit letter appears as Exhibit 1 — Jack Wireman (Part 3).

Opinions of the Geyser Pump Company

Geyser Pump Company engineers said they had in the past experienced somewhat similar wear problems in bearings of pumps like that being used with the Task motor. Their answer had been to use a custom bearing of higher angular contact designed by another bearing company. At the suggestion of Geyser Pump, Task contacted the company that had made the special bearing. The bearing company suggested for Task a custom bearing of higher contact angle similar to the custom bearing proposed by Orbit and costing around $1,000 each.

Opinions of Thomas Barish

The first two bearings, which had failed, together with prints of the motor and pump as shown in Jack Wireman (Part 1), were sent to Mr. Barish. His analysis of these bearings, which was dated December 14, 1963, and appears as Exhibit 2 — Jack Wireman (Part 3), was more extensive than that of Orbit. He, too, suggested a number of possible causes of failure. Rather than lubricity, Mr. Barish saw the main cause of failure as being fatigue, but he also suggested several other possible sources of difficulty. Among the suggestions in his report were modifications to the design of the pump as well as use of heavier bearings.

Mr. Barish recommended that special bearings be avoided and standard bearings used. With standard bearings, he said, there was less likelihood of unforeseen problems.

Meanwhile, more motors were being made and tested. The second qualification-test motor was shut down and disassembled after 630 hr, when metal particles began showing up in the discharge fluid. Already the front bearing, an Orbit No. 204HJB1519, was failing seriously. Pictures of this bearing after failure appear in Exhibit 3 — Jack Wireman (Part 3). With failure of this bearing at 630 hr, Mr. Wireman thought the evidence more strongly than ever suggested that heavier load capacity was not the answer, but he did not know how to explain the phenomenon or what to do about it.

116

Discussion Questions

1 What analytical techniques can be applied to un-
   ravel the mysteries of these bearing failures?
   Which, if any, can you apply, and what can you
   show with them?

2 Contrast the approach of the bearing company
   with that of Mr. Barish in trying to help Mr.
   Wireman. What light has each shed on the
   problem?

3 What should Jack Wireman do now, and why
   should he do it?

Exhibit 1 — Jack Wireman (Part 3)

ORBIT BEARING CORPORATION

December 4, 1963

Mr. Jack Wireman
Task Corporation
1009 East Vermont
Anaheim, California

Reference: My letter of September 11, 1963 to Mr. Elmer Ward

Dear Jack:

The bearings from the hydraulic pump which you sent for analysis after 700
hours of test running have been examined. The 204 size bearing is still in
serviceable condition in spite of some dirt denting including a few rather large
depressions caused by hard contamination in the .−2/.005 size range. The
general quantity of dirt denting, however, is less than that in the first bearing
we reported on last September.

The 204HJB1519 bearing shows evidence of advanced fatigue. There are
two types of fatigue patterns, the more prominent, which is fairly deep spalling,
extends for approximately one half of the circumference of the inner ring
contact area. There are also, in the remaining contact area, several small spots
of surface fatigue. In addition, there is appreciable dirt denting which most
probably resulted from the material plucked out of the raceway. There is no
evidence of excessive radial unbalance or thrust loading or of improper
mounting.

In comparing this bearing which failed at 700 hours with the original
204SST5 bearing which failed at 1800 hours, we find that this bearing has light
traces of surface fatigue and moderately heavy spalling in the raceway while
the earlier bearing showed extensive metal erosion from surface fatigue and
resultant heavy wear in the bearing. In other words, in this bearing the distress
that has occurred happened quite swiftly while that in the earlier bearing was
the result of a much slower process over a very long time. It is possible, indeed
probable, that the surface fatigue in the 204SST5 bearing had begun even
before the 700 hour point was reached. We cannot explain why the surface
fatigue in the 204HJB1519 bearing generated the heavy spalling while that in
the other bearing continued as a surface fatigue with general wear resulting.
It is, however, quite apparent that the heavy fatigue does not have the same
pattern as a typical, heavy load, subsurface fatigue. The fatigued ring was
checked for hardness with readings of 60.75 to 61 Rc with a specification
tolerance of 58.5 to 62 Rc.

The corrective action to take to overcome the failure condition is not
changed — to increase the capacity of the bearing so that the unit stress will be
decreased to a point where the lubricating qualities of the hydraulic fluid will
be adequate to sustain life for the required 3000 hours. In accordance with
our recent telephone conversations we have designed a bearing within the

envelope dimensions of the 204 bearing which we feel should be adequate for the application. This is the 204HJBX31 bearing. In this bearing the ball complement has been increased to ten 11/32 balls and the radial play has been increased to give a nominal contact angle of 35°. Inasmuch as the retainers in both of the above bearings were in practically new condition, there has been no change made in the retainer. The one being used is, again, 80-10-10 bronze. The increased ball complement and high radial play result in a design life of 500,000 hours at 75 lbs. thrust load. Bearings can be modified to this design in four weeks at a price of $175.00 each for six bearings or $60.00 each for 25 bearings. Prices for production quantities will be forwarded as soon as an inquiry can be processed.

The bearings are being returned herewith. Should you have any question on the above or if we can be of further assistance please don't hesitate to contact us.

Yours very truly,
THE ORBIT CORPORATION

Herbert D. Williams
Senior Product Engineer

HDW:etk
CC: Carl Berg, Purchasing Agent

Exhibit 2 — Jack Wireman (Part 3)

For Task Corporation, 1009 East Vermont Ave., Anaheim, California

Attention: Jack Wireman, Project Engineer

Pump-Motor 56383: Failure of 204 Ball Bearing.
(1) *General Conclusions:* Bearings failed because of excessive thrust load: plus minor contribution from fluid not being the best lubricant;
(2) *Detail Bearing examination,* pages 1 and 2. First pump end bearing too far gone to draw conclusions, except primarily inner race fatigue failure: cage behaved very well. No heating from rub or bad lubrication.

Second pump bearing showed clear fatigue failure, primarily thrust load, on inner. May have been appreciable radial load.

Both small end bearings showed much larger thrust load than expected 3 to 7 lb. from spring. Areas showed 92 lb. thrust (page 1) but this is misleading as it resulted from 2 or more load paths overlapping which looked like one large load. Nevertheless the actual load was 5 to 10 times the expected: indicating outer race did not slide in aluminum housing in operation (or spring was bottomed in error).
(3) *Expected Loads:* page 5; Pump spring, 40 to 70 (on test, 47)

Booster Pump	36
Spline Friction	*58 to 64*
Total Thrust	*134 to 190 lb.*

If the *spline imbeds,* the thrust can be *very much higher.* With the present design, imbedding with a sharp edge is almost certain because
(a) the unit pressure is considerable, 2190 psi projected area,
(b) the spring takes out all initial shake, and holds spline in one place with all motion at pump (none in motor),
(c) female spline at pump too soft.
In addition, there may be considerable axial expansions due to
(a) alum. housing and steel shaft at lower or higher (same temp.),
(b) shaft temporarily much hotter than housing, and

(c)   any lift-off at valve plate in pump, when operating.

Radial loads can come from misalignment, since long heavily loaded splines will have very little rocking effect. If assume spline tight, and max. runout of .004″, will give 31 lb. on motor bearing.

Also magnetic loads may be appreciable. However, failures indicate primarily thrust loads.

(4)   *Measured Loads* from contact areas on the bearings, page 6 show:

169 to 197 lb. thrust on second 204 bearing

but these figures are not exact: estimate −0 to +50%.

Thomas Barish
December 14, 1963

(5)   *Estimated Life:* Using latest ABEC formulas, second bearing (Orbit 204 HJB) has a rating of 473 lb. radial at 6000 rpm, (B10 life of 500 hours.) At 183 lb. thrust load, this gives *4200 hours* B10 life.

Apparently, it took appreciably more thrust load, + some radial, + inferior lubricity to bring this down to 700 hrs.

(6)   Even a loose bearing would not give appreciable radial motion to permit large magnetic loads. With thrust down to 80 lbs., max. radial motion is under .0002″ at 120 lb. radial. Curves, page 7.

Recommendations:

(1)   Check unit for possible *axial bind.* The second bearing, 204HJB showed a fairly constant heavy load, whereas spline thrust would vary.

Back check: thru any available aperture (or make one), push shaft or rotor against spring. Should move .010 to .020, after reaching 40 to 70 lbs., without much load increase.

(2)   Eliminate possibility of *spline imbedding:* For changeover that does not require any major parts change (except maybe truing motor shaft bore) see design 2, page 8. Need only replace quill by 2 piece design, and new longer spring. For new motors, recommend design 3.

(3)   *Increase thrust capacity* by changing to 40° or 35° bearing. These are in regular production by MRC (7204P); SKF (7204B, maybe); New Departure (30204); and Fafnir (7204W). Be sure to obtain either bakelite-compound or bronze one-piece cages: one with lots of space to allow oil (and foreign matter) to pass thru easily.

It is not necessary to have these made by others as specials, at greatly increased cost, and with some experimentation involved. The first two makes above have 10 to 15% more ball capacity, (34 to 54% more life.)

(4)   On new designs, cut diameter at booster pump next to bearing, and leave some debris space around bearing. See design 3, page 8.

(5)   Improve shaft fit for smaller (203) bearing. One bearing returned showed turning and loose fit. Needs tighter fit when no locknut. Use mfr standards, and be sure bearing slightly loose to allow for press fit. Recommend mfrrs "loose fit" standard.

(6)   Believe steel liner in housing will be necessary for small bearing. In spite of many efforts, no one seems to have been able to make this work without a sleeve for long life and any but very small loads.

In the meantime, for current units, recommend .0003″ loose min.

(7)   Review airgap selection; Petroff's equation shows about 3 h.p. loss in gap even at 2 centifposes viscosity and .007″ gap. At larger gap, less fluid loss may offset magnetic losses.

Loss Calculations:

Thrust Load:

(1)   Preload Spring on Quill: 40 to 70 lbs.
Actual on Test Units 47 lb.

Thomas Barish
December 9, 1963

(2)   Friction from Spline:
   11.5 H.P. $\times$ 63000/6000 rpm = 120.7 in. lb.
   Tangential force, 120.7/.23″ radius — 524 lb.
   Axial Friction at .11 to .16 coef. of friction
      = 58 to 84 lb.
   This number may be exceeded if spline tends to imbed.
   Spline Loading: Area: .46″ p.d. $\times$ .60″ long = .28 sq. in.
      (tangential projected area)
   Unit Loading: 524/.28 = *2190 psi:*
      (if equally divided.)

(3)   Thrust from Booster Pump:
   2.8″ O.D. at 6000 rpm = 73.3 ft./sec.
   Hydraulic head at 80% eff. $V^2 \times .8/2g$ = 67 ft. head
                        = 16 psi
   Thrust on seal diam., 1.7″   = 36 lb.

Total Thrust: 134 lb. to 190 lb. with possible large increase if
                        spline imbeds.
Radial Load: ½ rotor weight + 1 G unbalance = 10 lbs.
   Possible load from eccentricity if splines imbed:
   (or from pump housing tilt)
   For .004″ eccentricity (.002″ radius)
   on beam 2½ long $\times$ .34″ diam. (ends fixed) 31 lb.

Load Calculation from Brg Contact *Measurements:*

	First Set — 1800 Hrs.		*Second Set — 700 Hrs.*	
Orbit Brg. No.	204SS-TX5	203SS	204HJB1519	203SS
Balls	8	8-17/64	11-5/16	8-17/64

Data from Orbit

Curvatures %	52–52		52–52 =	52–54 52–52
Inner Groove Rad.	.1625		.1382	.1625
Groove Diam.	1.0175		.8564	1.0175
Land Diam.	1.1300		.9520	1.1311
Groove Depth	.0562		.0478	.0568
Degrees	48.5		48.5	49.5
Radial Looseness	.0005–9		.0005–9	.0015–19
Initial Contact Angle	11.47–15.85		12.5–16.8	16.25–18.3
Hardness-Inner				60–75–61 C.

Measurements and *Calculations* (inner races)

Contact a	.050	.050	.06	.050
b	.060	.060	.050	.060
Load/Ball (from b)	46	43	55	46
a + ½b	.08	.08	.085	.08
in degrees	34.	28.6	30.	34.
Contact Angle	14.5	20.9	19.0	14.5
Thrust Load	92 lb.*	169 lb.	197 lb.	92 lb.*

Contact angle, calculated using measured load and initial angle

T/nd²K	.00153	.00148	.000860	.00153
angle	18–20.4	18.5–21.2	19.6–21.3	18.5–21.2

(* both of these may have been two or more paths overlapped)

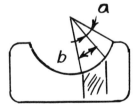

Thomas Barish
December 9, 1963

For Task Corp.
Motor-Pump Brgs.

T. Barish
Dec. 13, 63

Page 7

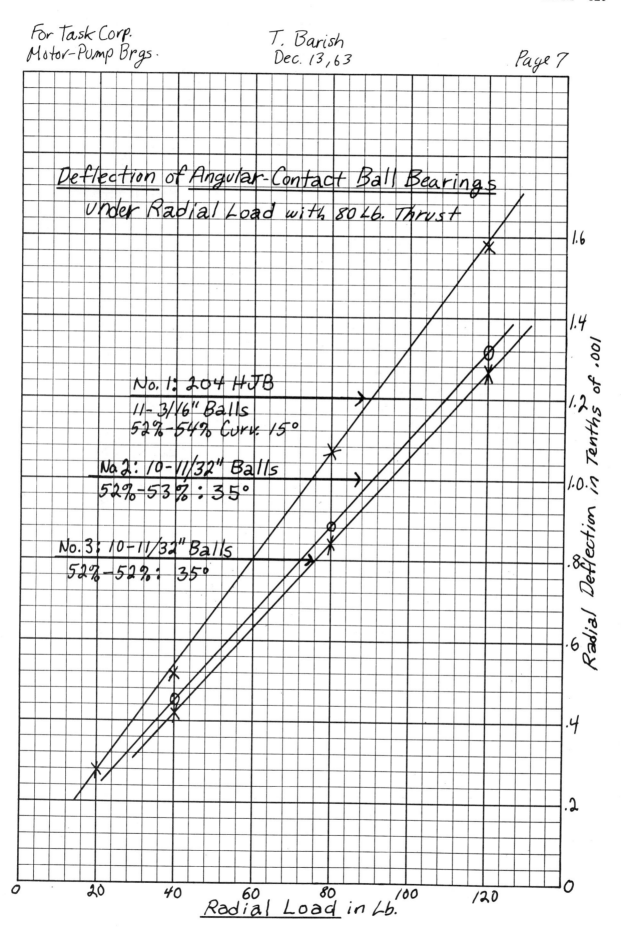

Deflection of Angular-Contact Ball Bearings
Under Radial Load with 80 Lb. Thrust

No. 1: 204 HJB
11- 3/16" Balls
52% - 54% Curv. 15°

No. 2: 10 - 11/32" Balls
52% - 53% : 35°

No. 3: 10 - 11/32" Balls
52% - 52% : 35°

Radial Deflection in Tenths of .001

1.6
1.4
1.2
1.0
.8
.6
.4
.2

0   20   40   60   80   100   120

Radial Load in Lb.

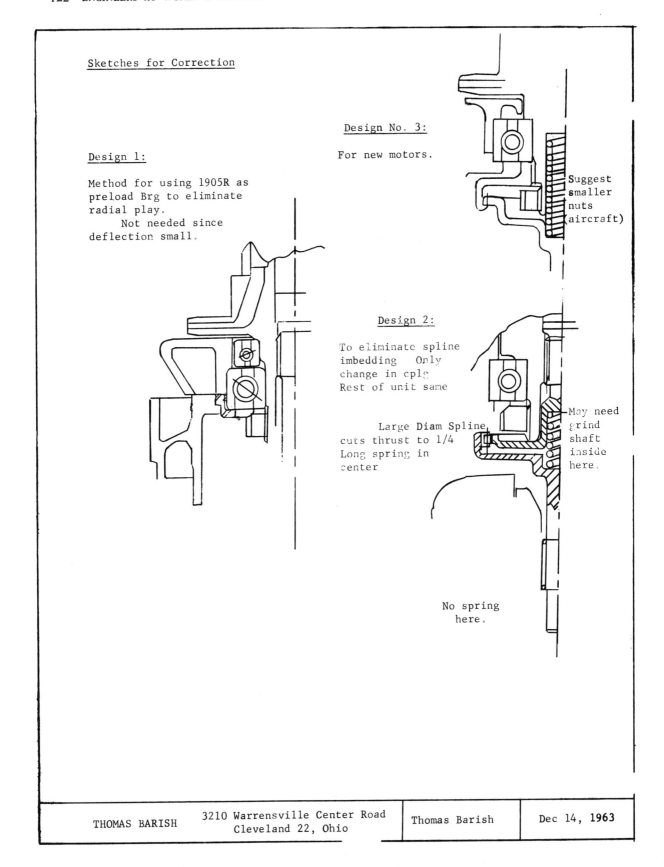

Sketches for Correction

Design 1:

Method for using 1905R as
preload Brg to eliminate
radial play.
    Not needed since
deflection small.

Design No. 3:

For new motors.

Suggest
smaller
nuts
(aircraft)

Design 2:

To eliminate spline
imbedding   Only
change in cplg
Rest of unit same

Large Diam Spline
cuts thrust to 1/4
Long spring in
center

May need
grind
shaft
inside
here.

No spring
here.

| THOMAS BARISH | 3210 Warrensville Center Road
Cleveland 22, Ohio | Thomas Barish | Dec 14, 1963 |

Exhibit 3 — Jack Wireman (Part 3)

Five DECISION MAKING

Making decisions is something anyone can do, but in doing so everyone makes mistakes. Disputes can be read in the press about which decisions are right and which are wrong at every level in politics. Scientific societies debate how to conduct their affairs, and in every debate there must be at least two viewpoints about every decision. Those who emerge as particularly effective in persuading others to adopt their decisions, and those whose decisions consistently tend to prove out well (these two not necessarily being the same), usually reach the ranks society rewards with exceptional pay, privilege, or both.

Some decisions are relatively easy to make and do not deserve much effort, such as which hand to hold the key in when opening a door. Others, such as whom and when to marry, generally deserve and receive much attention. Part of the art of effective decision making is being able to decide in advance which decisions deserve more attention. A humorous illustration is the cartoon showing the tail of a plane sinking in the middle of the ocean while nearby in a life raft the navigator says to the pilot, "But I was only off by one decimal point."

Some key facets of effective decision making, careful problem definition, generation of many alternative solutions, and use of analysis where appropriate to predict or optimize results have been raised in earlier chapters. The matter of choice of criteria was also raised in the discussion of an attribute matrix, which makes a grid to display alternatives against attributes that characterize them. To apply these effectively requires common-sense virtues, such as familiarity with the facts of each problem situation; understanding of what the physical evidence is, and what it means; objective and unprejudiced consideration of the facts; and separation of what is important from what is not. Formulas and rules do not help as much as simple practice does in developing these skills.

Formal procedures have been developed for treating some kinds of decision problems, such as rules of probability and expected value for playing games of chance where odds can be quantified. These are treated in books on mathematical decision theory, but they usually require oversimplification for ap-

plication to less-structured problems typical of real-life situations. Methods of estimating present value of costs and revenues receive considerable practical usage, particularly in banks and insurance companies, and these are explained in books on engineering economy. Books and articles in the area of mathematical psychology treat schemes for "weighting" criteria in assessing judgments, and describe how the human mind tends to operate in forming quantitative values for attributes normally considered subjectively. And finally, books on computer science describe ways that digital computers can be used to help with decisions ranging from choice of economic lot size to making medical diagnoses.

Forecasting of time and costs is an inescapable necessity in most engineering decision making, and a format used by the central engineering laboratory of one major company (FMC Corporation) appears in Fig. 5.1. Almost any kind of forecasting is a forbidding procedure because the forecaster is virtually bound to be wrong, and the only question is, By how much? By persevering through this discouragement and practicing, however, he will usually get better at being close, and his value will correspondingly increase.

The cases should therefore be used as dry runs to practice estimating, comparing results with others, and exploring the bases for differences in judgment, in order to strengthen this skill for a later time when it really counts on the job.

Other criteria generally vary from one situation to another, some possible examples being:

a   Weight (versus size, strength, capacity, and so forth).
b   Power consumption or output (versus efficiency or the above).
c   Life in cycles or years under various adverse conditions.
d   Noise and/or other emissions.
e   Appearance, convenience, comfort, safety.
f   Compatibility with other devices or materials.
g   Reliability/simplicity.

Typically, there will have to be tradeoffs between criteria such as these, and there will be room for judgment and debate about choices. A simple trick for displaying advantages and disadvantages of a particular alternative can be to run a vertical line down the center of a piece of paper (or blackboard), then list "pros" on one side of the line and "cons" on the other. (Usually pros of one alternative will be cons of another.) Listing them this way can help to get all the issues out in the open so they can be more intelligently and completely explored in a discussion without becoming focused on one and ignoring others of importance.

The result of decision making should be the development of a clear plan or set of recommendations. An easy mistake to make will be to stop short of being explicit about what those recommendations are, and to omit a description of specifically how the recommendations should be carried out. Who should put them into effect? What will he need to be able to do it? What should it cost in time and money, and if the resources to carry out your plan are to be diverted in some respect, where? Have you considered contingencies? What results do you expect from implementing your plan? What will you do if those results are not forthcoming?

Exercises

1 Fill in the development project worksheet cost estimation form (Fig. 5.1) with the best estimates you can for one of the cases studied so far or for one of the design concepts you generated earlier.

2 Recall six decisions you made, three of which led to distinctly favorable outcomes and three of which led to distinctly unfavorable outcomes. Describe any lessons you learned from those experiments in decision making.

3 Ask a practicing engineer to recall decisions and outcomes as in 2, above, and discuss what his experiences imply.

4 It has been a policy in your consulting engineering firm not to work on designing weapons systems, and in past surveys all your employees have agreed with this policy. Now the customer for whom you have been developing deep-sea-mining pressure transducers has asked you to incorporate modifications for use in torpedoes. He has made it clear that unless you undertake this additional task he will phase your firm out of the entire project, which would mean loss of financial support for three engineers, two technicians, and two secretaries on your staff. How would you go about deciding what action to take?

5 Discuss the statement, "If something I object to is to be designed and I'm not willing to do it, somebody else will."

6 Your firm is developing a water purifier for a chemical plant. You think you see a way to adapt some of the principles to a purifier for underdeveloped countries, but the project would cost many times the amount of money you could rasie. Your boss says the profit potential is not sufficient to justify such an investment by the company. Develop two alternative plans of action for yourself. List the pros and cons of each, stating your main assumptions. What would you do?

DEVELOPMENT PROJECT WORK SHEET

fmc CORPORATION

Original_____
Revision No.____
To Close_____ Date_____

PROJECT
X X X X X

Title_____

STEP	TYPICAL CONSIDERATIONS	MAN MO.	LABOR	BURDEN ___%	MATERIAL AND OUTSIDE CHARGES	G&A ___%	ESTIMATED TOTAL COST	WHOLE COST			ESTIMATED COMPLETION DATE
								SPENT TO DATE	APPROP TO DATE	APPROP TO BE REQUESTED	
1	**RESEARCH INVESTIGATIONS AND ENGINEERING ANALYSIS** a. EVALUATION OF REQUIREMENTS AND POTENTIAL MARKET. 　1. CONSULTATION WITH SALES AND MARKETING. 　2. CONSULTATION WITH POTENTIAL CUSTOMERS. b. PRIOR TECHNICAL ART 　1. PATENTS — STATE OF THE ART SEARCH. 　2. COMPETITION. c. CONCEPTION AND EVALUATION OF ALTERNATIVE SCHEMES. d. ESTIMATE OF TIME AND COST FOR ALL STEPS AND SPECIAL TOOLING. e. ESTIMATE OF MANUFACTURING COST. f. POSSIBLE FINANCIAL RETURN. g. PREPARATION OF REPORT TO INCLUDE: 　1. ANALYSIS OF CRITICAL TECHNICAL ELEMENTS. 　2. PERFORMANCE REQUIREMENTS. 　3. PRELIMINARY DESIGN SPECIFICATIONS. 　4. DETAILED RECOMMENDATIONS INCLUDING APPROPRIATION REQUEST FOR SUBSEQUENT STEPS IF NECESSARY.	Objectives: Supported By:						Work Performed By:				
	Total for Step 1										
2	**DEVELOPMENT EXPLORATION** a. THOROUGH STUDY OF ALTERNATIVES. 　1. CALCULATIONS. 　2. STUDY LAYOUTS AND SCHEMATICS. 　3. CONCLUSIONS. b. JURY-RIG TESTS OF UNTRIED PRINCIPLES. c. PATENT SEARCHES. 　1. PATENTABILITY. 　2. INFRINGEMENT. d. CONSULTATIONS WITH UNBIASED TECHNICAL SOURCES. e. CONSULTATION WITH MANUFACTURING AND SALES. f. PREPARATION OF REPORT TO INCLUDE: 　1. SUMMARY OF TEST RESULTS. 　2. MANUFACTURING FEASIBILITY OF SCHEME CHOSEN. 　3. REVIEW OF SALES DEPARTMENT COMMENTS. 　4. REVISED ESTIMATE OF TOTAL STEPS COST 　5. REVISED ESTIMATE OF MANUFACTURING COST. 　6. EVALUATION OF POTENTIAL SUCCESS. 　　(a) COMMERCIALLY (ECONOMICS). 　　(b) TECHNICALLY. 　7. RECOMMENDATIONS AND APPROPRIATION REQUEST FOR FURTHER WORK, IF NECESSARY.	Objectives: Supported By:						Work Performed By:				
	Total for Step 2										
3	**DEVELOPMENT MODEL DESIGN** a. PREPARATION OF LAYOUTS. b. RE-CHECK WITH MANUFACTURING AND SALES. c. CONSULTATION WITH OUTSIDE SPECIALISTS. d. PREPARATION OF DETAIL DRAWINGS. e. RE-CHECK PATENTS. f. ISSUE DRAWINGS FOR DEVELOPMENT MODEL CONSTRUCTION. g. PREPARATION OF REPORT TO INCLUDE: 　1. EVALUATION OF PROGRESS TO DATE. 　2. RE-EVALUATION OF ALL COSTS. 　3. DISCUSSION OF NECESSARY COMPROMISES. 　4. RECOMMENDATIONS FOR CHANGES TO SPECIFICATIONS. 　5. RECOMMENDATIONS FOR FUTURE ACTION AND NEW APPROPRIATION REQUEST, IF NECESSARY.	Objectives: Supported By:						Work Performed By:				
	Total for Step 3										
4	**DEVELOPMENT MODEL CONSTRUCTION** a. PURCHASE MATERIALS. b. METHODIZE. c. MANUFACTURE AND SHOP TEST. d. MODIFICATIONS. e. ENGINEERING LIAISON WITH SHOP. f. PREPARATION OF REPORT TO INCLUDE: 　1. MANUFACTURING FEASIBILITY. 　2. RECOMMENDATIONS FOR MODIFICATIONS TO EASE MANUFACTURING PROBLEMS. 　3. REVIEW OF MANUFACTURING COSTS AND PROJECTS COST TO DATE VERSUS ESTIMATE. 　4. RECOMMENDATIONS FOR FUTURE ACTION AND NEW APPROPRIATION REQUEST, IF NECESSARY.	Objectives: Supported By:						Work Performed By:				
	Total for Step 4										

form CEL 202 REV 1/68

STEP	TYPICAL CONSIDERATIONS	MAN MO.	LABOR	BURDEN ___%	MATERIAL AND OUTSIDE CHARGES	G&A ___%	ESTIMATED TOTAL COST	WHOLE COST			ESTIMATED COMPLETION DATE
								SPENT TO DATE	APPROP TO DATE	APPROP TO BE REQUESTED	
5	**DEVELOPMENT MODEL TEST** a. ESTABLISH TEST SPECIFICATIONS COMPATIBLE WITH PERFORMANCE REQUIREMENTS. b. INSTALLATION AND OPERATION OF MODEL. c. INSTALLATION OF INSTRUMENTATION. d. EVALUATION OF TEST DATA. e. SALES DEPARTMENT OPINION. f. PREPARATION OF REPORT TO INCLUDE: 1. COMPARISON OF TEST DATA WITH SPECIFICATIONS. 2. RECOMMENDED DESIGN CHANGES. 3. REVISED SPECIFICATIONS (IF NECESSARY) FOR PRODUCTION MODELS. 4. RE-EVALUATION OF ALL COSTS, COMPARING WITH ESTIMATES. 5. RE-EVALUATION OF COMPETITION. 6. RE-EVALUATION OF SUCCESS (a) COMMERCIALLY (ECONOMICS) (b) TECHNICALLY. 7. RECOMMENDATIONS FOR FUTURE ACTION AND NEW APPROPRIATION REQUEST, IF NECESSARY.	Objectives: Supported By:						Work Performed By:			
	Total for Step 5										
6	**PRODUCTION MODEL DESIGN** a. CONSULTATION WITH DEVELOPMENT ENGINEERS. b. CONSULTATION WITH OUTSIDE SPECIALISTS. c. RE-AFFIRMATION THAT DEVICE WHICH HAS BEEN DEVELOPED TO DATE WILL MEET SPECIFICATIONS. d. PREPARATION OF LAYOUTS, DETAILS, ASSEMBLIES, BILLS OF MATERIAL. e. SET UP LIAISON WITH DEVELOPMENT ENGINEER FOR HIS CONFIRMATION THAT FINAL DRAWINGS WILL MEET PERFORMANCE SPECIFICATIONS. f. RE-CHECK PATENTS. g. COST REDUCTION STUDIES. h. MANUFACTURING AND TOOLING COST ESTIMATES. i. RE-CHECK WITH MARKETING AND SALES. j. PREPARATION OF REPORT TO INCLUDE: 1. DISCUSSION OF HOW WELL DESIGN WILL MEET SPECIFICATIONS. 2. PREDICTION OF MANUFACTURING DIFFICULTIES. 3. RECOMMENDATION FOR AREAS OF FUTURE COST REDUCTION EFFORT AND/OR CHANGES TO IMPROVE SALEABILITY. 4. OTHER RECOMMENDATIONS FOR FUTURE ACTION INCLUDING APPROPRIATION REQUEST, IF NECESSARY.	Objectives: Supported By:						Work Performed By:			
	Total for Step 6										
7	**PRODUCTION MODEL CONSTRUCTION** a. TEMPORARY TOOL DESIGN AND MANUFACTURE. (IF PRODUCTION TYPE REQUIRED APPLY TO STEP 10.) b. PURCHASE MATERIALS. c. METHODIZE. d. MANUFACTURE AND SHOP TEST. e. MODIFICATIONS. f. ENGINEERING LIAISON. g. PREPARATION OF REPORT TO INCLUDE: 1. ACTUAL MANUFACTURING AND TOOLING COSTS COMPARED TO ESTIMATES. 2. EVALUATION OF MANUFACTURING PROBLEMS. 3. RECOMMENDATIONS FOR IMPROVEMENT.	Objectives: Supported By:						Work Performed By:			
	Total for Step 7										
8	**PRODUCTION MODEL TEST** a. ESTABLISH TEST PROCEDURE TO BE COMPATIBLE WITH PERFORMANCE REQUIREMENTS. b. CHECK WITH DEVELOPMENT ENGINEER WHO TESTED PREVIOUS MODEL FOR COMMON PROBLEMS. c. INSTALLATION AND OPERATION. d. EVALUATION OF TEST DATA. e. IMPARTIAL EVALUATION PREFERABLY BY COMPETENT TECHNICAL MAN NOT CONNECTED WITH ORGANIZATION. f. CUSTOMER REACTION — SALES DEPARTMENT? g. PREPARATION OF REPORT TO INCLUDE: 1. COMPARISON OF TEST RESULTS WITH PERFORMANCE SPECIFICATIONS. 2. RECOMMENDED MODIFICATIONS FOR FIRST PRODUCTION LOT. 3. RECOMMENDED FINAL INSPECTIONS. 4. REVIEW OF COMPETITION. 5. REVIEW AND RE-EVALUATION OF SUCCESS. (a) COMMERCIALLY (ECONOMICS) (b) TECHNICALLY 6. RECOMMENDATIONS FOR FUTURE ACTION AND APPROPRIATION REQUEST, IF NECESSARY.	Objectives: Supported By:						Work Performed By:			
	Total for Step 8										
9	**FINAL ENGINEERING DRAWING** a. FINALIZE DRAWINGS AND BILLS OF MATERIAL. b. SPECIAL ENGINEERING-MANUFACTURING INSTRUCTIONS. c. PREPARATION OF DATA FOR INSTALLING, OPERATING AND SERVICING. d. TRAINING SERVICE PERSONNEL. e. RELEASE OF DRAWING. f. PREPARATION OF FINAL REPORT TO INCLUDE: 1. REVIEW OF ALL COSTS COMPARED TO ESTIMATES. 2. FINAL REVIEW OF ECONOMICS WITH CURRENT PREDICTION FOR SALES OBTAINED FROM SALES DEPARTMENT. 3. REVIEW OF ENGINEERING PERFORMANCE, NOTING AREAS WHICH SHOULD BE IMPROVED. 4. FURTHER RECOMMENDATIONS.	Objectives: Supported By:						Work Performed By:			
	Total for Step 9										
	Total for All Steps										

form CEL 202 REV/1-68

HENDRIK VAN ARK

Request for a Bridge Design

Most highway bridge designs come from handbooks, but how? Where does judgment enter in this case, and which bridge type should Hendrik choose?

"Every so often there's a rush job that has you working through Saturdays," Hendrik Van Ark observed. As a civil engineer in the Structural Division of the Lancer Construction Company, he had been called to a meeting on Thursday, April 13, 1967, concerning a new, two-unit, 1,500-Mwatt steam power plant being built by Lancer 2,000 miles away on the Missouri River. The question the group had for him was how to bridge a creek lying between the construction site and a county road that presumably would be used in trucking men and equipment to the site. An independent consulting engineer, Mr. Brynne, had recommended building a 100-ft, composite-steel-girder bridge because, he said, it would probably be possible to obtain the girders quickly. Mr. Van Ark was asked to consider the recommendation and proceed as he believed appropriate. "Normally I wouldn't be called into a meeting like that where everyone else was a consultant or chief engineer of some department, but this bridge design was needed fast to avoid delaying other work."

A plan indicating routing for the new access road to the site was available (Exhibit 1 — Hendrik Van Ark), and from conversation at the meeting Mr. Van Ark had learned further facts about the job. He was told that heavy pieces of machinery to be brought to the site would include a rotor, weighing roughly 100 to 150 tons; stator, weighing 200 tons; and casings and steam drum, which might weigh from 50 to 80 tons each.

Men and equipment were at present reaching the site by dirt roads and small farm bridges. But for hauling larger loads of cement and other materials these were considered totally inadequate. The standard two-lane county road, which ran with its centerline parallel to and approximately 20-ft from the edge of the creek, was being resurfaced to enable it to support the largest conventional diesel truck trailers, one of which might weigh as much as 72,000 lb fully loaded, be 8-ft wide and 50-ft long, and have a minimum turning radius of 40 to 50 ft. The new

bridge was being built to allow such trucks to reach the power-plant site from this county road.

"The Missouri State Highway Code doesn't say much about the maximum load that can be carried by truck," Mr. Van Ark continued. "But the AASHO (American Association of State Highway Officials) code recommends a limit of *20* tons. State highway officials can impose limitations sometimes even when a code does not, so I think it pays to be conservative. In this local area of Missouri, for instance, the judge who issues all building permits 'strongly suggests' that we build a bridge, rather than simply running the road across fill with culverts in it for the water to run through. A bridge makes more sense for flooding conditions anyway, but they tell me the judge's reason is partly that he just likes the looks of a span better.

"An advantage of fill embankment would be that it would probably only cost around $40,000, but the judge isn't much concerned about that. Actually, in this particular situation the customer isn't either (although he can't admit it) because he's a public utility. The government allows them to earn profits as a percentage rate of return on their investment, so the higher the investment, the higher the profit they can take. Maintenance costs, on the other hand, *are* considered important to this utility, according to our group supervisor who went down there and talked to them."

Mr. Van Ark inquired about flooding conditions, and was told that the water level rose to within a foot or two of the top of the banks (to the 483-ft elevation mark — see Exhibit 2 — Hendrik Van Ark) every year or two. "I asked for photos of the banks," he said. "If they appear to be all eroded it's a sign there is lots of flooding, whereas if they are covered with brush there's probably not much flooding to worry about. If there is a flood only every 25 yr you don't have to be so conservative in the design. Here the pictures suggested fairly frequent flooding. It shouldn't do any serious damage to have water running over the bridge once in a great while, but naturally it would be better not to."

He also asked for confirmation of the creek cross-section drawing done by people at the site, and for soil borings in the banks down to rock for later use in designing abutments to which the span would be mounted. An earlier cross-section drawing had been made for general layout of the plant road, but it did not show the precise location of the creek bed relative to the county road.

His next task, as he saw it, was to do a brief analysis for deciding which of several types of bridges would be best. He thought the total cost of the bridge, which he expected would be about $100,000, did not justify an extensive study, but he pointed out that unless some study was used there might be an uneconomic choice of alternative that would later reflect poorly upon Lancer as a construction company, and damage its reputation. He noted that because the bridge was intended for use in construction that had already begun, it was desirable that his design be done quickly.

Hendrik was in the middle of other work when he was called in to design the bridge, and it was not until Saturday that he had enough free time to begin. He started with some rough computations to estimate costs for several alternative types of bridges, including the one recommended by Mr. Brynne. "Once the location is decided, the first thing to do is figure out what type of bridge it should be," he observed. "There are several possible steel bridges, and there are also some using prestressed concrete girders. I asked about the location of shops for making girders, and it turned out that there were shops within 500 miles for making either steel girders or concrete ones. I phoned one of them in Omaha, Nebraska, and the representative said he could guarantee delivery within a month.

"Mr. Brynne thought steel girders would be best. He's an authority — particularly on bridges and on wind-loading of structures — who has many years of experience with these things. He comes in every Thursday and discusses whatever problems we bring up. I have experience, too, but mine is more in structures for nuclear power plants. I don't pretend to be a super-duper specialist, but just design for whatever comes along. There's more variety that way. On bridges I generally like to get Mr. Brynne's ideas first, then check them out and do my own thinking from there. Some of the other engineers prefer to do their own thinking first and then check with him. It's mostly a matter of taste.

"There are generally two types of concrete girders, the pretensioned kind and the posttensioned. The pretensioned type is completely prepared in the shop for installation, whereas with the posttensioned you have to do the tensioning in the field after it is in place. When the site is too far from a shop or when the girders have to be too long or heavy to transport, it's better to cast them in place and then tension them. But this takes more time, so I think we should only consider pretensioned girders if we use concrete. An advantage of concrete girders is that they should be cheaper to maintain, because they don't have to be painted every 2 or 3 yr."

In comparing concrete and steel girders Mr. Van Ark saw a number of alternatives open, and his first step toward making the choice was to draw from the Lancer Company library a copy of the "Manual of Bridge Design Practice" (prepared by the State of California Division of Highways).* "Handbooks are essential in this business," he commented, "so you don't spend all your time reinventing the wheel. Lancer Corporation has a collection of its own data on various topics, and generally each public agency has one or more manuals giving its own particular set of rules and standards. They may overlap to some extent, but each will usually have its peculiarities. For instance, the highway engineers are normally more conservative than building engineers, and the railroad engineers are more conservative than the highway engineers, maybe because the things they build don't go out of style so fast and have to last longer.

"Manuals are especially useful for designing fairly standard structures like this bridge. Here we have parallel sides, square ends, and a moderate length, so the handbook fits nicely. If I wanted to put in a curve or some unusual feature, the handbook wouldn't be enough. I'd have to solve problems of loading and strength using basic theory. But here I only need enough theory to know that the handbook gives the right information for this job." †

Mr. Van Ark began his computations with assumptions listed in his notes, which have been copied as Exhibit 3 — Hendrik Van Ark. Since the county road had two lanes, engineers at the meeting had said the bridge should, too, and he estimated a deck width (rail-to-rail) of 30 ft. The standard load-carrying ability (denoted by H20-S16-44) he described as the heaviest one listed, and from experience he recalled that it was often used for roads leading to power plants. According to the bridge-design manual:

The H20-S16-44 load represents an approximation of a train which consists of a 20 ton truck preceded and followed by 15 ton trucks.

Hendrik considered again the question of whether to use steel or prestressed concrete girders. Another

---

* The Preface explains, "The graduate engineer is expected to be well grounded in the fundamentals of structural design, but until he gains practical design experience, he may encounter difficulties in incorporating this fundamental knowledge into efficient and rational design procedures. The manual is intended primarily for the inexperienced bridge designer, but the older or more experienced engineer may find new ideas and methods which would be of value to him."

† Mr. Van Ark's education had included a degree in Civil Engineering from the Netherlands Academy of Arts and Technological College in Rotterdam. He worked as a structural design engineer for 6 yr before joining Lancer and had been with Lancer for 5 yr as of 1967.

question was whether to use one span or three. He noted that freeway overpasses often used two spans with a pier in the middle because shorter spans could be lighter and cheaper and easier to erect. He rejected the possibility of using two spans across the creek, however, because of the difficulty and expense of constructing a pier in the center, where there was always water. Whether the three spans should be separate or one continuous beam laid across intermediate piers was a question he chose to ignore for the moment, and to consider later only if the choice between one span and three were close. He thought a single beam laid across the piers, rather than three separate beams, might reduce by about 10 percent the amount of steel required. He also thought that in his first comparative cost estimate he would ignore costs of excavation and backfill needed for abutments at the ends of the bridge, of decking, and of transporting materials to the construction site. These expenses he expected would be roughly the same for any alternative.

His computations of girder costs for several alternatives are shown in Exhibit 3 — Hendrik Van Ark. Charts from the bridge-design manual on which they were based appear in Exhibit 4 — Hendrik Van Ark. A summary of his computations appears on the first page of his notes, with blank spaces for the figures he thought should be calculated next. To compute the cost of abutments and piers he planned to use the creek profile drawing (Exhibit 2 — Hendrik Van Ark) and the manual charts (Exhibit 5 — Hendrik Van Ark). He thought a cantilever type of abutment would be suitable for the ends of the span, and that it would be reasonable to assume that the center piers would be similar to hinged abutments, in making computations for the three-span design. The abutment height ("H" on the chart of Exhibit 5 — Hendrik Van Ark), he explained, should be measured up from a point roughly 3 or 4 ft below the surface of the ground. Below this point would be pilings on which the abutment rested.

"I think it's important to be meticulous about calculations like this," he said, "because you never know when you may have to go back and check how you worked something out. The company regulations say all calculations are to be filed permanently and that we are supposed to keep them neat, but I'll bet less than 25 percent of the engineers do it. I've seen some of them squirm in meetings when somebody asked them what assumptions they had used to get a certain figure, and they couldn't remember. My calculations have to be neat anyway or I can't keep my thinking straight.

"Once it is clear which type of bridge we should have I'll go on with more details of the design, such as how many piles to use for the abutments, and what pattern to place them in. It will probably take me 1 or 2 weeks to prepare a complete design we can send out to subcontractors for bids. By then I'll have to work out about 40 to 50 pages of calculations, at least two drawings, plus a collection of other paperwork such as standard specifications, purchase orders, memos, and correspondence. Hopefully, construction on this bridge will begin by the end of May at the latest."

Discussion Questions

1. Carry forward the work on Hendrik's project as far as you can with the time and case information you have.
2. Describe the process by which you defined problems and made decisions in this project. What aspects would you most like to do more checking on?
3. Identify the "people problems" in this project and comment on how they should be handled, both ideally and practically.
4. Comment on the decision-making procedures in this case as opposed to one or more of the other cases you have studied.

Exhibit 1 — Hendrik Van Ark

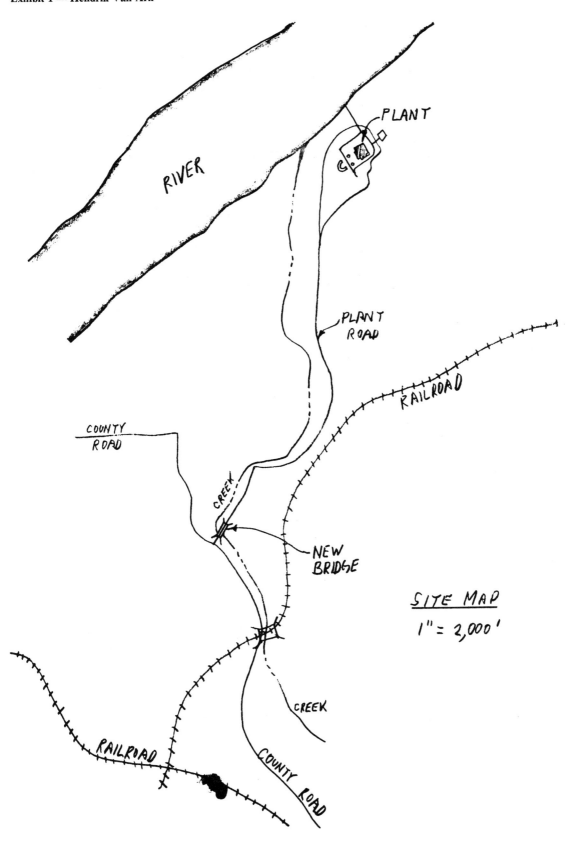

Exhibit 2 — Hendrik Van Ark

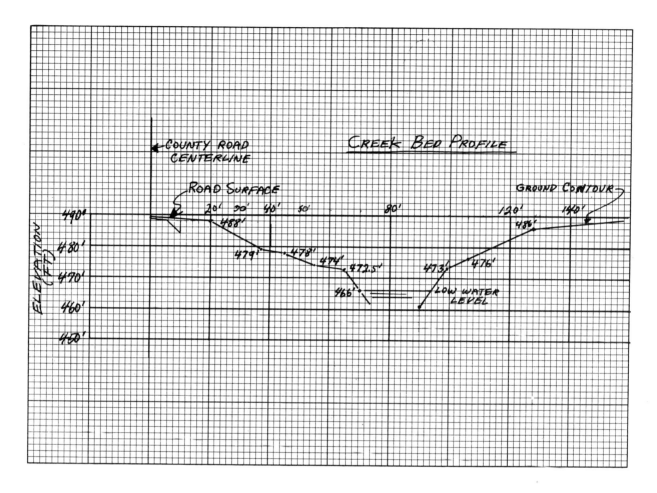

Exhibit 3 — Hendrik Van Ark

4-15-67

BRIDGE TYPE - SELECTION STUDY - BASED ON FIRST CONSTRUCTION COST

REFERENCE - MANUAL OF BRIDGE DESIGN PRACTICE
STATE OF CALIFORNIA - DIVISION OF HIGHWAYS

ASSUMPTIONS - 1. DECK WIDTH = 30'. STANDARD LOADING AS GIVEN BY MANUAL IS H-20-S16.

2. GIRDERS TO BE MADE IN SHOP AND CARRIED TO SITE.

3. SO STRUCTURAL STEEL OR PRESTRESSED CONCRETE GIRDERS CAN BE CONSIDERED.

4. 1 SIMPLE 100' SPAN IS ADEQUATE; TO BE COMPARED WITH 3-35' SPANS. COST DECREASE FOR POSSIBLE CONTINUITY WILL NOT BE REFLECTED.

5. COST OF GIRDERS AND ABUTMENTS WILL BE CONSIDERED ONLY.

6. ABUTMENT WILL BE CANTILEVER TYPE.

7. PIERS WILL BE HINGED TYPE.

COST SUMMARY

BRIDGE - TYPE SUPERSTRUCTURE	100' SPANS			3 - 35' SPANS		
	GIRDERS	ABUTMENTS	TOTAL COST	GIRDERS	ABUTMENTS & PIERS	TOTAL COST
ROLLED STEEL BEAMS				$13,000		
COMPOSITE GIRDER 36 W F SERIES				$10,800		
COMPOSITE WELDED STEEL GIRDERS	$21,600			$10,800		
PRESTRESSED CONCRETE "I" GIRDER	$9,810					

4-15-67

COST CALCULATIONS - GIRDERS

1. 100' - SIMPLE SPAN - COMPOSITE WELDED STEEL GIRDER

THIS TYPE USED FOR SPANS OF 60' TO 140' LENGTH. (FROM CHART)
GIRDER SPACING 7'-6" (FROM CHART)
STRUCTURAL STEEL WEIGHT (FROM CHART) 30 P.S.F.
DECK AREA = 30 FT × 100 FT = 3,000 S.F.
WEIGHT OF STEEL REQ'D = 30 PSF × 3,000 S.F. = 90,000 #
* COST @ $480/TON IS 90,000 × $\frac{480}{2000}$ = $21,600 WORTH OF STEEL,

2. 100' - SIMPLE SPAN - PRESTRESSED CONCRETE "I" GIRDER

USED FOR SPANS 30' TO 120'. GIRDER SPACING 7'-0
DEPTH 5 FT.   4 GIRDERS REQUIRED.
QUANTITIES FROM MANUAL CHARTS
*   PRESTRESSING STEEL   4 (1,800 #) = 7,200 # @ 55¢/# = $3,950
      REINFORCING BAR   4 (1,800 #) = 7,200 # @ 12¢/# = 860
      CONCRETE   4 (450 FT^3) = 1,800 FT^3 = 67 CUBIC
                       YARDS @ $75/CY   = 5,000

    4 GIRDERS   TOTAL COST $ 9,810   WORTH OF CONCRETE AND STEEL

3. 3-35' SIMPLE SPANS - ROLLED BEAMS

USED FOR SPANS 25' TO 75'. GIRDER SPACING 7'-6'
STRUCT. STEEL WT. FROM CHART 18 PSF
TIMES DECK AREA 3,000 SF = 54,000 #
* COST @ $480/TON = 54,000 × $\frac{480}{2000}$ = $13,000 WORTH OF STEEL

4. 3-35' SPANS - COMPOSITE GIRDER - 36 WF SERIES

USED FOR SPANS 50' TO 80'. GIRDER SPACING 7'-6"
STRUCT. STEEL WT., SAY 15 PSF.
TIMES AREA 3,000 SF = 45,000 #
* COST @ $480/TON = 45,000 × $\frac{480}{2,000}$ = $10,800 WORTH OF STEEL

---

* Cost Figures were from company records. (footnote added by casewriter.)

4-15-67

<u>Cost Calculations - Girders (Cont.)</u>

<u>5.</u> 3-35' Spans - <u>Composite Welded Steel Girders</u>
Used for spans 55' to 140'. Girder Spacing 7'-6" to 10'-6"
Struct. Steel - Although for a 35' span there would be
a slight decrease in weight per sq. ft., this
probably would be offset by an increase in
fabrication labor cost. So assume cost of
struct. steel is same as for case 4. above.
= <u>#10,800</u>

<u>6.</u>   3-35' Spans - <u>Prestressed Concrete "I" Girders.</u>

Exhibit 4 — Hendrik Van Ark

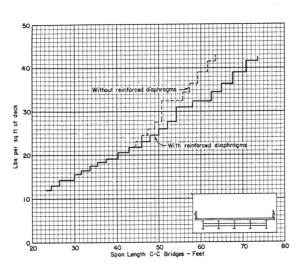

Notes : Quantities based on 7'-6" girder spacing.
Weights of diaphragms included.

Loading : H 20 - S16 - 44   Reinforcing steel : f_S = 20,000 psi
Structural steel : f = 18,000 psi   Concrete : f_C = 1,000 psi

STRUCTURAL STEEL
SIMPLE SPAN ROLLED BEAM SUPERSTRUCTURE

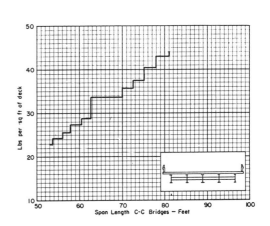

Notes : Quantities based on 7'-6" girder spacing.
Weights of diaphragms and shear connectors included.

Loading : H 20 - S16 - 44   Reinforcing steel : f_S = 20,000 psi
Structural steel : f = 18,000 psi   Concrete : f_C = 1,000 psi

STRUCTURAL STEEL
SIMPLE SPAN COMPOSITE GIRDER SUPERSTRUCTURE
36 WF SERIES

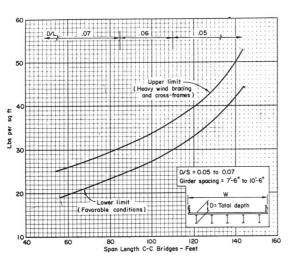

Notes : Weights of diaphragms, stiffeners, shear connectors, and minimum of wind
bracing included.

D/L ratio approximates maximum economy

Loading : H 20 - S16 - 44   Reinforcing steel : f_S = 20,000 psi
Structural steel : f = 20,000 psi   Concrete : f_C = 1,000 psi

STRUCTURAL STEEL
SIMPLE SPAN COMPOSITE GIRDER SUPERSTRUCTURE
WELDED GIRDERS

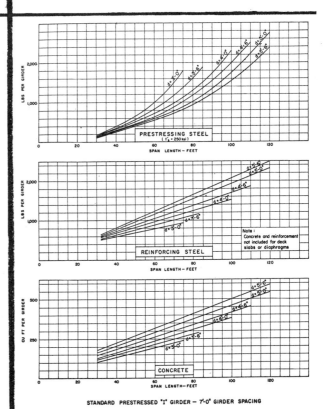

STANDARD PRESTRESSED "I" GIRDER — 7'-0" GIRDER SPACING

Exhibit 5 — Hendrik Van Ark

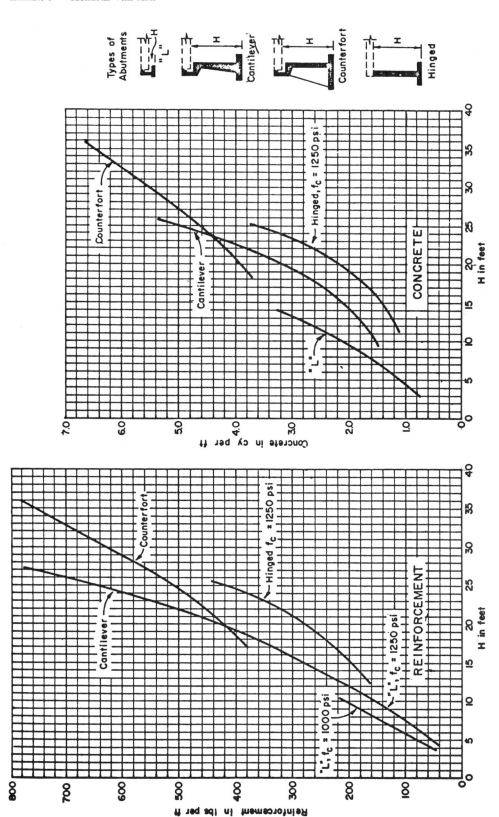

ABUTMENTS

LABORATORY EQUIPMENT CORPORATION*

Machine-drawing Usage

How does this company machine its products with no use of machine drawings? Should they continue to do so? What are drawings really for anyway if things can be made without them?

In the spring of 1956, Harold Quillian, the vice president of Laboratory Equipment Corporation of America, was considering hiring a part-time draftsman to prepare shop-drawings of the numerous parts involved in the manufacture of the approximately 100 items in the company's product line. Use of such drawings would represent a change in the practices the organization had followed for more than a half-century. Mr. Quillian was therefore anxious that such a move be made only if it would result in significant advantages to the company.

Laboratory Equipment Corporation was located in Mason, Connecticut, near a large metropolitan center. Although small in size, the company enjoyed a reputation as one of the world's outstanding manufacturers of laboratory equipment for teaching and research in the field of physiology. The firm's principal customers were secondary schools, universities, medical colleges, hospitals, and research laboratories, both in the United States and abroad.

The company had been established as a single proprietorship in 1904 by a doctor who was well-known as a teacher and researcher in physiology. At the time of its founding, physiological laboratory equipment was produced almost exclusively by German manufacturers and available in the United States

* This case, Laboratory Equipment Corporation of America, EA-P 214, was prepared as the basis for class discussion rather than to illustrate either effective or ineffective administration.
Copyright © 1956 by the President and Fellows of Harvard College. Used by specific permission.
This case also appears in *Casebooks in Production Management: Basic Problems, Concepts and Techniques,* Revised Edition, by Arch R. Dooley, William K. Holstein, James L. McKenney, Richard S. Rosenbloom, C. Wickham Skinner, and Philip H. Thurston (New York: John Wiley & Son, Inc., 1968).

only at considerable cost. As a result, medical schools and universities were restricted in their ability to offer physiological laboratory courses that demanded large quantities of equipment for student use. Research in the field was similarly hampered by the high cost of equipment. In an effort to overcome this situation, the doctor had established a small manufacturing organization and begun to produce, at reasonable cost, equipment for use in his own teaching and research activities. Word of this spread quickly among professional circles, and the organization was soon besieged with requests to manufacture similar equipment for sale to schools and laboratories. As a result, the doctor, although retaining his faculty affiliation, formed a small company to produce such items.

The company's sales volume grew steadily over the years, and more than doubled during the period following World War II. This reflected the large increases in college enrollments and in the volume of research activity throughout the world at that time. Foreign orders, particularly, increased during this period, due in large measure to financial grants by private foundations to support medical research and education in underdeveloped regions of the world.

The year 1955 brought the largest sales volume in company history, and indications were that 1956 would equal or exceed this level of activity. This current volume approached the limits of company capacity with existing facilities and manpower. If this rate of growth continued, it would soon be necessary for the company to consider such alternatives as physical expansion, multishift operation, or a large increase in the use of subcontractors.

The current product line consisted of approximately 100 items. Changes in the line were infrequent, and the majority of items had been manufactured in their present form for many years. Items were rarely dropped from the line, even when their typical annual sales volume had fallen to modest levels.

Retention of such slow-moving items reflected both the company's desire to render service to the profession and the characteristics of the demand for certain of its products. In the case of many types of basic laboratory equipment for the classroom, rugged designs had been developed that would stand up under student use over a period of years. Most of the schools needing such equipment had, in prior years, already purchased sufficient quantities to meet normal student requirements for a long period of time. Thus, even though thousands of such items might still be in active use throughout the world, new orders coming into the company for these slow-moving products were small during any given year. Inevitably, however, a few orders would be received to replace units that had become lost or damaged beyond repair. In the interest of classroom uniformity, purchasers usually desired that the replacements be exact duplicates of the items purchased in prior years. To satisfy these needs, it was necessary for Laboratory Equipment Corporation to continue such items in their product line, even though their annual sales of these items were modest.

Similar conditions existed with regard to replacement parts. It was company policy to offer repair service on all items it sold, without any time limit. Since a small but steady volume of service orders was constantly received, it was necessary for the company to continue to produce and stock replacement parts for products for which the volume of new sales was almost negligible.

From time to time new items were added to the company's line when new inventions were perfected or new research techniques were adopted in the field of physiology. In the past the development of such new products had been infrequent, and at most had involved only a few items during any single year. New developments were taking place in the field of physiology, however, particularly in research techniques. Mr. Quillian therefore believed that opportunities for new products, especially those involving electronic applications, were considerable, and that the company would probably accelerate the launching of new items in the future. Since the firm's capacity was almost entirely absorbed in the production of the established product line, in recent years it had been necessary to subcontract the manufacture of any new products that were added. For at least the immediate future, Mr. Quillian believed that it would be necessary to continue this practice because, in most instances, new items supplemented rather than replaced existing products in the line.

The regularly manufactured items varied as to size and complexity. Some were relatively simple and posed few manufacturing problems. An example was stainless-steel probing instruments used for dissections. Other products, such as recorders, which were used to record and measure slight variations in pressure or in nerve impulses, were complicated pieces of equipment consisting of numerous parts and requiring a high order of manufacturing skill.

The company's shop area was equipped with the basic machine tools commonly employed in metal-working operations. The company also maintained its own plating department and a large assembly area. All production activities were under the supervision of the production manager, E. F. Bush. In the machine shop Mr. Bush had the assistance of a working foreman.

None of the firm's 15 production employees was specialized, but instead each was capable of performing a number of different types of manufacturing operations. When parts were machined in small lots it was not uncommon for a single operator to perform a number of different process steps requiring the setup and operation of several types of machine tools.

In none of its manufacturing activities did the company employ shop drawings. Instead, an alternative system had been developed over the years to preserve and convey information about each of the numerous manufactured parts. Immediately adjacent to the shop area, a series of shelves had been constructed to a height of approximately 10 ft along a wall. The shelves were 8 in. apart, and each of them contained a row of small wooden boxes, approximately 5 by 6 by 15 in. in size. Each box contained all of the tools and fixtures required to manufacture one particular part, together with actual samples of that part as it appeared at each of the successive stages of its manufacture. These items had originally been prepared, through the combined efforts of a designer and various shop personnel, when an item was first added to the product line. The designer would describe what was required, and the shop personnel would make experimental models until a satisfactory prototype of the part was obtained. If special tools or fixtures were required, these were developed experimentally at that time, with the assistance of a tool and die maker. The completed items would then be placed in the box for future use. Exhibit 1 — Laboratory Equipment Corporation shows one of the boxes holding its complete contents. Exhibit 2 — Laboratory Equipment Corporation shows the contents removed from the box.

Each box bore a label identifying its contents. Other than for the practice of placing less-frequently used boxes on the upper shelves, the location of the boxes did not follow any regular pattern. From experience and repeated use, however, the machine-shop foreman knew the location of each of the boxes and could usually locate any particular one of them in a matter of seconds.

An example of the use of the boxes may be illustrated by reference to a part whose manufacture required several machining operations on a casting purchased from an outside foundry. If the required operations consisted, in sequence, of (1) a turning and facing operation on an engine lathe, (2) the milling of two slots on a milling machine, and (3) the drilling of two small holes on a drill press, the box for that part would contain each of the following:

1 a sample of the casting as originally received from the foundry;
2 the facing and turning tools required for the lathe operation;
3 a sample of a casting after completion of the lathe operations;
4 the milling cutters required to mill the two slots;
5 a sample of a casting after completion of the lathe operations *and* the milling operation;
6 the drills needed for the holes;
7 a sample of a casting after completion of *all* the machining operations, including the two drilled holes;
8 any jigs or fixtures needed at any of the manufacturing steps.

When a lot of this particular part was scheduled into production, the machine shop foreman would go to the shelf area, locate and pull down the box containing the objects previously mentioned. In some of the boxes, the various samples were mounted on cards, in proper sequential order. In other boxes the contents were mixed together in a wholly random fashion. In this event it was necessary for the foreman, drawing upon his personal knowledge and experience, to sort out and arrange the items to show, step by step, the various manufacturing stages.

By use of the box's contents, the foreman could provide an operator with actual samples of the part at each process stage, and with the tools and fixtures to be used in its production. These samples could be picked up and examined by the operator. Any dimensions that were in doubt could be confirmed by actual measurements.

After the operator had completed all the manufacturing operations on an item, it was his responsibility to replace the contents of the box and to return it to the foreman. The foreman would, in turn, replace the box on the shelf, making a mental note of its position so he would be able to relocate it.

Since most of the machine-shop operators were old-timers with many years of company service, there were numerous occasions when it was not necessary to refer to the boxes unless they contained special tools or fixtures not otherwise available in the shop. Instead the operator would know from memory what operations were required, and could use the shop's supply of general-purpose tools.

Although no accurate records had been kept on the subject, Mr. Quillian estimated that 40 or 50 percent of the boxes saw regular use one or more times each year. Another 15 or 20 percent were probably used no more than once every 2 yr. The remainder, in his opinion, contained inactive items that were scheduled into production very infrequently. As a result, these boxes were, at most, probably used once every 3 or 4 yr.

Mr. Quillian, who had been with the company since 1953, knew that this particular method of preserving information about manufactured parts differed from that used by most manufacturers. Firms doing subcontracting work for the company often asked for blueprints, and in such cases it was necessary for the company's designer to prepare rough sketches showing shape, dimensions, and so on. These, plus verbal instructions from the designer, had usually enabled the subcontractor to proceed satisfactorily. Mr. Quillian knew, however, that in at least some instances, subcontractors had seen fit to prepare conventional blueprints of the item for use in their own shops.

On balance it appeared that the company's system had worked reasonably well for more than 50 yr. Mr. Quillian was therefore undecided about whether there would be any real advantages in converting to the more conventional use of shop drawings. He had discussed the matter with both the shop superintendent and the machine-shop foreman. Neither of them had indicated any strong reaction either in favor of or in opposition to the idea.

While considering the matter, Mr. Quillian had conducted exploratory conversations with an experienced draftsman who would be available for part-time work in the near future. The man's qualifications were such that Mr. Quillian was sure he could prepare accurate drawings from the items contained in each of the boxes. For such work the draftsman would receive $2/hr and could be available for approximately 10 hr per week. Suitable space could easily be arranged for the draftsman in the company offices, and Mr. Quillian did not feel that any close supervision of his activities would be necessary.

Since there were approximately 350 boxes on the shelves, Mr. Quillian felt that this was a reasonable estimate of the number of drawings that would have to be prepared if it were decided to make a complete conversion to their use. The wide variations in the complexity of parts made it difficult, however, to estimate the total hours of a draftsman's services that would be required for such an undertaking. In discussing the matter, the draftsman stated that he believed it would require about 2 hr to prepare a drawing of a "simple" part, such as a double clamp. This would include careful lettering of all written material, exact measurement of all drawings to scale, accurate spacing of the various views upon the page, and other steps required to produce a finished drawing. If Mr. Quillian would be satisfied with rough drawings, requiring less care in preparation but showing the same information, the draftsman estimated that the time could be cut by perhaps one-third. Parts more complicated than the double clamp would, of course, require proportionately more time for either rough or finished drawings. Like Mr. Quillian, the draftsman found it difficult to estimate the total workload that would be involved to prepare drawings for all of the parts in the product line.

Mr. Quillian knew that before joining the company most of the present operators had used shop drawings, either while employed by other firms or while students in trade schools. Since the majority of these men had been with Laboratory Equipment Corporation for at least 10 yr, however, Mr. Quillian found it difficult to estimate how much of this skill had been forgotten through disuse and whether refresher training would be necessary. Three of the men had joined the company as apprentices, and had had no other form of training or experience. For such men, the introduction of shop drawings would definitely require a certain amount of preliminary instruction.

Discussion Questions

1 What do you think the Laboratory Equipment Corporation should do about its problem? Explain your reasoning. What factors might reasonably occur that would be sufficient to change your decision?

2 Comment on this statement: "The main reason this company should change to engineering drawings is that it is generally a standard practice to use drawings."

3 Comment on this statement: "The company's system has always been the way it is and it has worked, so they should not make the change-over."

Exhibit 1 — Laboratory Equipment Corporation

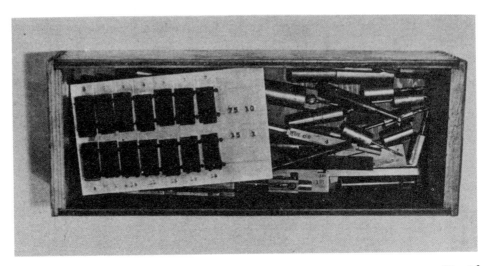

Box and its Contents Immediately After Removal From Shelf

Exhibit 2 — Laboratory Equipment Corporation

Contents Themselves

The REVEREND C. W. VAN DOLSEN

Electrical Appliances Overseas

What should this man, who lacks engineering expertise, do about using his electrical appliances overseas? He asked some consultants. But they seem to disagree.

In September, 1965, a letter was received by DATA International of Palo Alto, California, from a missionary who was preparing to go to Uruguay, asking advice about electrical appliances he could purchase and use on the 60-cycle voltage supply in the United States as well as on the 50-cycle voltage supply in Uruguay.

Background

Development and Technical Assistance (DATA) International was organized in 1959 as a volunteer organization to operate as a clearinghouse for technical information for Americans living and working in remote corners of the world. These people frequently found themselves with technical problems they could not solve wihtout help. DATA began by specializing largely in helping American missionaries.

With the advent of the Peace Corps and the increasing numbers of voluntary-agency representatives abroad, DATA was branching out to offer its services worldwide to all Americans working at or near the village level.

Since its inception DATA had received more than 5,000 letters from people all over the world, each asking help with one or more problems. DATA had enlisted more than 1,500 specialists — engineers, scientists, doctors, and related technicians — willing to give time and effort without pay toward solving these problems.

When a request for help was received, DATA customarily sent the problem to at least three of its consultants to obtain independent opinions. They also sent inquiries to manufacturers. Replies from these sources were then sent to the person requesting help.

Such a request was received from C. W. Van Dolsen of Warrensburg, Missouri, who was about to leave for Uruguay as a missionary. A copy of his letter is shown in Exhibit 1 — The Reverend C. W. Van Dolsen.

Following the customary procedure, DATA forwarded copies of the letter with a request for assistance to consultants, whose replies are shown as Exhibits 2 to 5 — The Reverend C. W. Van Dolsen. These replies were forwarded to the Reverend Mr. Van Dolsen. His question was then what action to take.

Discussion Questions

1 To what extent do the Reverend Mr. Van Dolsen's advisors agree or disagree? How do you explain their agreement or disagreement?
2 What should the Reverend Mr. Van Dolsen do about his problem?
3 To what extent do you think engineers should contribute their energies to an organization such as DATA International without charging money? Explain your reasoning and comment on the extent to which you think medical doctors and lawyers do or should reason the same way.

Exhibit 1 — The Reverend C. W. Van Dolsen

Cozetta, C. W. and Warren

MISSIONARIES TO URUGUAY,

SOUTH AMERICA

4399

C. W. Van Dolsen
703 South College
Warrensburg, Mo.
September 30, 1965

Dear Sir:

We have been referred to you by Missionary Equipment Service
of Chicago. In going out to Uruguay I have some questions
concerning the electric current. Uruguay requires appliances
to be used before they enter the country. This means that
we must use them here and there. Realizing that their current
is 220 volt 50 cycle, I am wondering how to do this.

The refrigerator seems to have no problems. I can buy one
with a motor for 110 volts 50/60 cycles. I can use it here
and with a transformer to change their 220 to 110, I can use
it there.

Is there an electric range on the market that has the 50/60
arrangement for the clock, timer, etc? Our electric ranges
are 220 volt. Will this work with their current. It has
been brought to my attention that our 220 volt require three
wires and perhps theirs only requires two. Is this correct?
And, would it make a difference other than perhaps a different
type plug? Are these change over plugs available?

I can buy an automatic washer either 220 volt 50 cycle or 110
volt 60 cycle. Could I just buy the 110-60 to use here and
change the motor for the 220-50 for use there? How much change
would there be? By changing the motor and perhaps a pulley
and belt, would this make the complete change so the the washing
cycles etc. would work properly?

I would also need to know the same information for an electric
dryer. Or, could I buy a 220 volt 60 cycle dryer here and use
it there expecting the time of the drying cycles to be off somewhat?

Thank you for any information you can give me.

Sincerely,

C.W.Van Dolsen

C. W. Van Dolsen

UNDER APPOINTMENT BY THE FOREIGN MISSIONS DEPARTMENT   •   ASSEMBLIES OF GOD
1445 BOONVILLE AVENUE   •   SPRINGFIELD, MISSOURI

Exhibit 2 — The Reverend C. W. Van Dolsen

international

437 CALIFORNIA AVENUE
PALO ALTO, CALIFORNIA

Answer Form

PROBLEM NO. _____4399_____

YOUR ANSWER:

| please type or print
ON THIS SIDE ONLY |

To be of maximum value a reply
should be received by _27 Oct_ 65

You should have no trouble operating a 110 volt 60 cycle
refrigerator in Uruguay providing, as you suggest, you use a
transformer to step voltage down from 220 to 110. The 50 cycle
current would only cause the unit to run at 5/6 the speed it
should run on 60 cycles, but this should cause no difficulty in
a refrigerator.

The U. S. standard for electric ranges is 220/110 volts,
3 wire. This means that between two "line" wires the voltage
is 220 volts and between either "line" wire and "neutral"
the voltage is 110. Standard U. S. ranges are designed to
utilize 110 volts on the low-medium heat positions. I am sure
Uruguay uses only 2 wire 220 volts which is not suitable for a
standard U. S. made range. It would be possible to get a
transformer for this application, but it would be very costly
as ranges consume a considerable amount of power. The timer
and clocks would run at 5/6 the normal speed anyway since the
frequency would be 50 cycles.

The same comments on electric ranges would apply to the electric
dryer.

It would be feasible to use a transformer to step voltage
down from 220 to 110 volts for a washer as they don't consume
much more power than a refrigerator. Of course the motor and
timer would also run at 5/6 the speed as on 60 cycles. This
would only mean, for example, with the timer set at 20 minutes,
it would would actually run for about 24 minutes. Of course
the machine will run at slightly slower speed, but this
should not be harmful.

Some major manufacturers of appliances in this country make
appliances that will operate at other voltages, such as 220
volts, 50 cycles. I would suggest you write to Westinghouse
International Company, 200 Park Avenue, New York, or General
Electric International Company in New York City, or some other
such as Frigidaire. I am sure one of these companies can
arrange to sell you appliances with proper voltage and frequency.
However, it is questionable if you would be able to use them
here before you left. I would suggest following the manufacturers'
recommendation on this point.

DATA International, on behalf of the Staff, Consultants, Sponsors and Assistance Agencies co-
operating, is pleased to render this problem-solving service, but must of necessity disclaim all
liability for its use or application.

DATE _____

FROM _____

James R Monroe

Exhibit 3 — The Reverend C. W. Van Dolsen

![DATA international logo] **international**

437 CALIFORNIA AVENUE
PALO ALTO, CALIFORNIA

OCT 1 8 1965

Answer Form

PROBLEM NO. _____4399_____

YOUR ANSWER:

To be of maximum value a reply
should be received by Oct 27, 65

please type or print
ON THIS SIDE ONLY

This problem concerns use of 60 cycle A. C. electrical
equipment with 50 cycles or 50 cycle equipment with 60.   The answers
are complicated because alternating current impedance (a kind of
"resistance" which holds back current and prevents too much flowing
through the equipment, which otherwise might overheat and burn out)
is proportional to the number of cycles per second.   In 1934 I took
my General Electric Refrigerator to Sumatra where they also had 220
volts but 50 cycles.   I bought an auto transformer here to reduce
the voltage from 220 to 110, but was advised by the shop foreman there
that a further reduction was necessary.   The amount of reduction may
be estimated by considering that most of the impedance of an electric
motor is attributable to the inductance of the windings (relatively
little resistance loss in the wiring).   If the motor is rated to stand
120 volts at 60 cycles, then five sixths of that voltage (50 divided
by 60) at 50 cycles will send the same amount of heating current through
its windings.   That means if your transformer would deliver 120 volts,
a further reduction of one sixth that amount or 20 volts is necessary.
A resistance capable of dropping voltage by 20 would produce heat and
cause excessive charges for unused or wasted electricity.   A choke
coil, on the other hand, dissipates the voltage but feeds the energy
back into the line so there would be only minor additional charge for
unused kilowatt hours of electricity.

The choke coil which I made consisted of a toroid (dough-
nut) of lengths of soft iron wire.   The larger diameter of the toroid
was about 8 inches, the smaller about 1 inch.   Around the toroid was

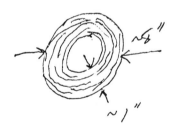

wound regular insulated electrical wire,
about number 12 or 14.   Caution: be
sure you have adequate electrical insulation to guard against shock hazard,
which is especially vicious with 220
volts and an auto-transformer!   The
actual amount of wire used had to be
arrived at by cut and try until the
voltage drop across the coil when connected in series with the terminals to
the icebox was enough to reduce that
across the icebox to the specified amount.

A different coil might have to be wound for each different
60 cycle motor you might want to use with 50 cycles.

Continued on following page.

DATE *Oct 15, 1965*   FROM *Donald S Villars*

Donald S Villars

Now, to discuss your specific questions. Unless your refrigerator motor for 50/60 cycles has different input connections for
the different number of cycles, I would be skeptical that it might overheat if you let it run on the slower number of cycles. (Also, the
motor is almost sure to run only five sixths as fast. In our case,
it was running 100% of the time in the heat of the tropics-- not cutting on and off like it normally does here at 60 cycles.)

I know of no electric range that has a 50/60 cycle clock or
timer. These are usually run by synchronous motors and if the number
of cycles is 50, they will only run five sixths as fast as at 60. (I
doubt that you will find any spring wound clocks or timers on an electric range.) The clocks probably also run on 110 volts. The heating
part of the range is completely independent of the number of cycles, so
no choke coil would be required. You will probably not find three
wires for 220 volts in Uruguay. (If you do you better get local instruction!) If you can identify which is the middle wire of your 3,
you can try leaving it unconnected, but beware of your clock and timer,
as they might get burned out if they work on 110 volts and depend upon
the two 110 v. sides being accurately balanced so as to give exactly
110 v. to the clock at <u>all</u> <u>settings</u> of your burner switches.

Changing a motor there to another which runs at 220 v., 50
cycles sounds expensive. Also you will not be able to claim it as
"used". Since you are buying new equipment and are not confronted
with the problem of having to make do with what you already have, <u>I</u>
would feel inclined to prefer to buy the 220 volt 50 cycle equipment
here plus a transformer from 110 to 220 (an auto-transformer used to be
cheaper) <u>for</u> <u>60</u> <u>cycles</u> here. The equipment might not run as well or
as efficiently here but I don't believe you would be likely to get into
as much trouble as you could by doing it in reverse fashion; i. e., try
to make do on 50 cycles with 60 cycle equipment and have to wind several choke coils.

I wish you luck in your tour of duty in Uruguay. Mrs. Villars and I spent a day in Montevideo on a tour of South America last
February. I was very glad to leave because it was the only country
we encountered with so many insulting anti- U. S. signs.

Exhibit 4 — The Reverend C. W. Van Dolsen

437 CALIFORNIA AVENUE
PALO ALTO, CALIFORNIA

Answer Form

PROBLEM NO. ___4399___

YOUR ANSWER:

> please type or print
> ON THIS SIDE ONLY

To be of maximum value a reply
should be received by _27 Oct 6_ 5

In my limited time on this business trip to the General Electric facilities at Valley Forge, Pa, I have gotten this information from my local inquiries & from a Field Engineer for International General Electric. This is all the assistance I can render immediately but perhaps it will help.

First the refrigerator. G.E. units sold in the United States are rated for either 115 volts at 60 cycles or 100 volts at 50 cycles. An electric motor runs hotter on 50 cycles and in a hermetically sealed refrigeration unit the voltage must be lowered to reduce heat dissipation within the unit. Since all refrigerators produced in the U.S. have similar units I am pretty sure that any refrigerator you take should be operated at 100 volts (plus or minus 10 percent, or 90 to 110 volts). Make sure that the transformer you buy has taps on it so you can select the optimum voltage output and make sure the transformer is big enough. For 1/4 horsepower motor use a transformer rated at 550 volt-amperes or more and for a 1/3 horsepower motor use at least a 750 VA transformer.

Next the electric range. There is no way to convert a clock or timer from 60 cycles to 50 cycles. Ranges in the US are wired for 3 wire 220 volts. One wire is neutral, one wire is plus 115 volts, and one wire is minus 115 volts. Both 115 volts and 230 volts is available within the stove. The clock, timer, lights and appliance outlets use this 115 volts. On stoves with 5 level heat control, 115 volts is used for 2 levels of heat. An electric stove with no clock, timer or flourescent lights and having either 3 level heat control or infinite heat control could be easily converted for 2 wire operation in Uruguay. Any incandescent bulbs (like the oven light) would have to be replaced with 220 volt bulbs. I am sure a different connector is required and is not available in the U.S. Uruguay has 2 wires, one neutral, and one 220 volts. On the stove,

DATE _11/1/65_   FROM *Robert H Johnson*
                                Robert H. Johnson

connect the neutral terminal and the negative terminal together and tie them to the neutral wire and to a convenient water pipe or metal stake driven into the ground. Tie the 220 volt wire to the plus terminal on the stove. There is one warning. An electric stove can draw at least 9000 watts (or 9 kilowatts, written 9kw). Be sure and check with the local authorities in Uruguay to see if 9kw is allowed on a 2 wire single phase line. They probably will not allow it or will require the installation of additional service.

Now the washer. On export models of GE washers, the 50 cycle models use a larger clutch drum and heavier shoes. The relay must match the motor current so the 50 cycle models have a different relay. These 3 parts must be changed to convert a 50 cycle model to a 60 cycle model. The heavier drum and shoes overload the motor at 60 cycles. To convert to 50 cycles from 60 cycles the motor must be changed also. The 60 cycle motor is not built to stand the extra heat dissipation. It was estimated that these three parts would cost approximately $15— and would take 2 hours to install. I did not get a price on motors. You probably cant buy a 50 cycle GE washer or parts. GE does not change the timer or gear ratios on their washers. They use the same for both 50 cycles and 60 cycles. But they do wind the timer motors with larger wire to handle the increased current on 50 cycles. If you can buy a washer for 220 volt, 50 cycle operation that does not require these changes in order to work on 60 cycles, you could just connect to the 230 volts, 60 cycles available here in the US and use it, as is. There is another factor to be considered also and that is water pressure. If the water pressure falls below 10 pounds per square inch, the valve on an automatic washer will not open. GE sells a semi-automatic washer that must be filled by hand. When you have dumped enough water into it, the wash (or the rinse) cycle will start. I do not believe any of the GE export models are available in the US.

And last, the drier. If you can buy a drier that is rated for both 50 cycles and 60 cycles, the motor would not have to be changed. However the drier, like the range, is supplied with 115 volts also and the timer motor is 115 volts. You would have to install a small transformer to supply 110 or 115 volts for the timer motor.

These answers are sketchy and quick. I already wish I had time to go back, rewrite, clarify and add more information. But possibly just this much will help you to ask more specific questions so that we can be of further help.

Yours in the Lord's service

Robert H Johnson

Exhibit 5 — The Reverend C. W. Van Dolsen

 **GENERAL ELECTRIC**

IGE EXPORT DIVISION

GENERAL ELECTRIC COMPANY
159 MADISON AVE., NEW YORK, N.Y. 10016
–U.S.A.–

AREA CODE 212
TELEPHONE: PLAZA 1-1311
CABLE ADDRESS: "INGECO, NEW YORK"

January 28, 1966
JWR - 2084

Mr. Douglas K. Hayward
DATA INTERNATIONAL ASSISTANCE CORPS.
437 California Avenue
Palo Alto, California 94306

Dear Sir:

With reference to your problem number 4399, please be advised as follows and within the limits of this problem as they apply to General Electric appliances only.
Due to design differences between manufacturers the information hereinafter provided
may not be taken as being applicable to major appliances manufactured by other
companies.

Our refrigerators and freezers, normally used on 115 volts, 60 cycles, can all be
used on 50 cycles if the voltage is transformer reduced to <u>100 volts</u>, 50 cycles.
In our case we do not like to see 110 volts still less 115 volts applied to the
motor when the line frequency is 50 cycles.

There is no such thing as an electric 50/60 cycle clock. It is either one frequency
or the other and mechanical parts have to be changed (rotor) to compensate for the
difference in frequency. If the voltage has to be changed either a small transformer
has to be used or the field coil has to be changed.

Most American electric ranges are designed for 236/118 volts, 60 cycles, 3-wire,
single phase service. This type of service normally provides 5 heats on the surface
units. A single phase 220 volt, 2-wire service only provides 3 heats and among
other components all the switches have to be changed and these ranges are specially
manufactured. Changing the plug does little or nothing.

The conversion of a standard domestic washing machine or dryer for use on 220 volts,
50 cycles involves changing many components. Compared with the first cost of a
machine manufactured for this type of electrical service, the conversion charge is
usually uneconomical. It should be realized that the washer motors work in conjunction
with a relay that is frequency sensitive and this also has to be changed.

In order to handle inquiries of this type we have set up five Authorized Exporters
in New York, Washington, D.C., Miami, San Francisco and Los Angeles. These outlets
carry the special export material and are equipped to take care of the needs of U.S.
continentals proceeding overseas. We attach a sheet showing the addresses of these
outlets.

Very truly yours,

J. W. Reynard
PRODUCT SERVICE

IGE-42B (2-64) REV.

JWR:cs

JACK WIREMAN

Failure of a Ball Bearing

Part 4

Jack called in a second consultant, who disagreed diametrically with the first, based on totally different lines of reasoning. Time and wherewithal are both short. The customer is very upset about delays. Now what?

Following failure of the third bearing at 630 hr, several further actions were taken by Task to determine the cause and cure. Additional bearings were sent to Mr. Barish, who examined them and supplemented his earlier reports with more possible explanations and reiteration of some of his earlier recommendations. A second consultant, Professor Dino Morelli of the California Institute of Technology, was asked to examine the bearings and give his opinions. Professor Morelli said he thought the cause was skidding of the balls due to too light an axial loading for the particular type of bearing when running fully immersed in fluid. The recommendations of Professor Morelli differed from those of Mr. Barish, leaving Mr. Wireman with the question of which of the recommendations should be followed, if any.

Further Opinions of Mr. Barish

As stated above, the third set of bearings was sent to Mr. Barish for examination. In addition, at his request, several sets were copper-plated and run for 10 hr, one set on a pump and one set on the Task Dynamometer with no pump, after which they were sent to Mr. Barish. From impressions on the copper, it could be seen where the balls were riding in the races and how wide the band of contact was, which in turn enabled him to compute axial shaft loadings.

A report on the third failure and an analysis of the copper-plated bearings were submitted by Mr. Barish on December 26. Some possible causes of failure suggested in this report were: (1) binding between housing and rear bearing and housing, and (2) axial vibrations due to the hydraulic pump.

From the copper-plated bearings, Mr. Barish concluded that axial forces on the shaft were within the predicted range, though the pump imposed a somewhat higher thrust than was to be expected from the loading spring alone. This report appears as Exhibit 1 — Jack Wireman (Part 4).

Another report by Mr. Barish, dated December 31, gives his findings for examination of five additional bearing sets that had run in service from 10 to 20 hr. In this analysis, which appears as Exhibit 2 — Jack Wireman (Part 4), Mr. Barish cited binding or misalignment of the rear bearing as the most likely source of difficulty.

Still later in correspondence, Mr. Barish made the following comments: "I apparently lost the needed emphasis on the major problems in trying to cover all possibilities and also because the first reports were made without seeing parts or bearings.

"Major Problems: friction in the spline produces thrust load, estimated at 54 to 68 lb. This always occurs, whether there is imbedding or not. Imbedding can cause much greater loading.

"The spline problem is aggravated because the spring held the 'floating' coupling tight one-way, and even small movements would start to build up thrust. Also the small diameter and greater length mean higher loads and little alignment capacity.

"Second: Usual practice uses steel liner in aluminum housings for ball bearing outer rings.

"Temperature changes *and* temperature differentials cause radial binding between outer ring and housing on the small bearing. Hence, there is much more thrust until the bearing is free to slide. Even transients here can be destructive."

A characteristic of the bearings that was discussed with Mr. Barish over the telephone was that of "ball banding," which had appeared on some of the service units' bearings. On some of the bearings, each ball had acquired a narrow circumferential stripe of wear, indicating that it had rolled in a continuously repetitive pattern. When some of the bearings had been taken for copper plating to the Bearing Inspection

Company — a specialized Los Angeles testing laboratory — Mr. Wireman had discussed the ball banding with representatives of the laboratory. They told him the banding showed that overloading of the bearing had prevented the balls from moving in a random fashion. Mr. Wireman asked Mr. Barish about this on the telephone, and was told that ball banding always occurred under operation with continuous thrust and was not an indication of overload.

Bearing Inspection also informed Mr. Wireman that bearing problems on pumps such as the one being used with the Task motor were not uncommon. In other pumps having the same general design and operational characteristics there had been problems of bearing failure. Mr. Wireman understood that the approach taken to solve such problems had been to install bearings of higher load capacity.

Opinions of Professor Morelli

Ball skidding due to a combination of factors was seen by Professor Morelli as the main cause of failure. These factors were as follows:

1   By operating fully immersed, a retarding force of viscous drag was imposed on the balls as they passed around the race. Catalogs on bearings, it was pointed out, specify "mist lubrications" for ball bearings. (In contrast to Professor Morelli, Mr. Barish saw full immersion of the bearings as an advantage. In his opinion, it assured full lubrication and the continuous flow tended to reduce the danger of localized heating.)

2   Being outer-race centered, there was an additional retarding force on the balls exerted by the retainer as it dragged through the viscous, shear against the stationary outer race. The point at which the retainer contacted each ball also caused it to apply a force inward on the balls, lowering the contact pressure between the balls and the outer race.

Professor Morelli concluded that the answer was to increase ball contact stress, to use an inner-race-centered bearing, or to modify the design so the bearings would not be running fully immersed. His recommendations were to use an inner-race-centered bearing of load capacity comparable to that of the bearing originally used.

Discussion Questions

1   What possible explanations you did not think of have been suggested by others in the case or in class to explain the bearing failures? How close did you come to thinking of them, and what additional mental programming would have carried you the rest of the way to thinking of them? How could you get such programming?

2   What alternative actions, constraints, and pressures should Wireman keep in mind as he decides what to do now? How should he make up his mind?

3   What action do you recommend to him? What should he plan to do if taking your advice does not solve the problem?

Exhibit 1 — Jack Wireman (Part 4)

for the *Task* Corporation, 1009 E. Vermont Ave., Anaheim Calif.
Attn. Jack Wireman, Project Engineer

Supplement A to Report of Dec. 14, 1963 on
Motor *Bearing Failures* in *Motor-Pump* 56383.

(1)   Examination of the new failure, SN/5, Orbit 204HJB (page A3) is
helpful because failure had not progressed so far. It showed that failure *started*
with the balls *riding on the edge* of the groove. The ball surface breaks were
typical.

The races did not have any real surface breaks, only severe roughening:
except outer did begin some slight surface breaks.

Severe roughening extended over entire grooves.

(2)   The balls could ride the edge of the inner under two conditions: the first
a very severe tilt of the outer, of the order of .005 to .010″/″. The outer race
surface roughening was over all the race except a few long islands in the center
(sketch, page A3). This also agrees with bad tilt. Likewise cage deterioration
with severe ball-pocket wear and light wear between cage and outer.

(3)   The second possibility (and I believe more likely) is that the outer race
of the small brg was bound in the aluminum housing and did not slide: happens
if aluminum housing is cooler than shaft and brg. Then also, the outer housing
would shrink endwise and bind small brg against large end brg. Practically
every small end brg showed much larger thrusts than spring preload. (109 lb.
this set, page A3)

On the first failure (conrad brg) this reverse load on the 204S would be less
harmful; than on later 204 hjb. Here, the reversing would quickly cause ball
to ride on shallow shoulder (about .008″ end movement) and this would break
up the ball surface rapidly.

The new failure rode on both edges of outer path. Brg would then break
loose and bounce back to other side: showing clear islands again. Also cage
could show the same distress, mostly from broken ball surfaces. Also inner
contact extended about 20° over-center towards wrong side.

(4)   Other possible causes:

Critical frequencies, axial: first using pump spring, = 1700 rp.m. Too low to
matter. Second using brg spring rate, pages A5 and A6 give critical of 38,000
per min. Still low, since one-per-cylinder forcing function is 54000.

Shock loads, when small brg lets go quickly, only rise to 660 lb.; small compared to shock capacity (brinell load) of over 2000.

Recommendations:

(1)   Same as before: check carefully for errors in handling, chips, binding,
etc. in assembling pump and motor.

(2)   As before: eliminate spline imbedding by two-piece coupling per sketches.
This will also isolate pump further from motor for eccentricities, transfer of
forcing vibrations, and will permit longer preload Spring.

(3)   Increase thrust capacity on 204 as before.

(4)   Add much emphasis on eliminating bind in 203S outer race fit. A good
steel liner, and looser fits are indicated. In view of trouble, recommend about
.0004″ min.

A good changeover is available with present parts by changing from 203 to
103 size leaving room for .10″ thick liner.

Plated Brg experiments proved very helpful: The contacts were all where they should be and showed no trouble (measurements and calculations, page A4) except that load on small brg was large as on all previous cases. Also the no. 3 brg (small end) shows a slight misalignment of outer race, and smaller load.

The positive conclusion that can be drawn from this is: these brgs would not have failed if continued as they were. The failures came from conditions introduced by some other operating conditions such as cold starts or unusual temperature differentials: or else by some assembly or machining errors existing on other units.

Examination of Additional *Failed Brgs:* S/N No. 5 Motor   600 Hrs.

(1)   Brg. Orbit 204HJB1519. This brg shed new light on the problem. Both races were fairly completely covered by very bad roughing up of the surface with only some parts of outer groove breaking thru the surface slightly. This existed over all the outer, and on inner from shoulder to about 20° over center. Both races had the ball run hard enough on the shoulder edge to raise a positive burr.

In addition the outer race showed some long islands, about at the middle of the roughened area, and about 1/5 of the width where the surface was relatively smooth. (sketch below)

Most of the balls were badly broken on the surface and some of them along a diam, about 180°, and much deeper than the race surfaces. This means that the balls rode the edge of the groove and the breaks were started by this edge cutting, (both on inner and outer).

This also means that the initial failure was on the ball surfaces from riding this edge and that the race roughening resulted from the ball failure.

In addition, the outer showed signs of having rotated slightly, on both O.D. and thrust shoulder.

The above condition, wide path with clear area in center of outer, is produced by (1)* a badly cocked outer race plus thrust load. Reference — "Effect of Misalignment on Forces Acting on Cage" by K. Kakuta, ASME Paper 63-Lub-12-.

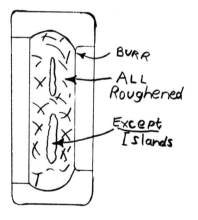

BURR

ALL Roughened

Except Islands

This was confirmed by the unusual retainer deterioration. All the ball-pocket surface showed bad spreading, some very severe with high burrs: whereas the O.D.-Outer-ring contact showed only a little distress. It was thought that unequal ball size might account for the variation in pocket wear but enough of the balls could be measured on unbroken diameters to show no wear (all .3125) and all equal.

Also the cage enlarged slightly under severe forces and bound slightly on outer ring.

Normally, such a tilted outer would show a bad tilt at the ball path, instead of the wide spread. But in this case, the outer turned somewhat in the housing.

There were no signs on inner or outer of appreciable chips under the thrust shoulders, or uneven clamping.

(2)   The small Brg: Orbit 203S showed no failure. Contact area measurements, a = .045": b = .06" (page 6, 1st report) which indicates 109 lb. thrust or two overlapping paths. Thrust was right way.

Bore showed no contact except one corner, and also appreciable rub at shoulder. Again, too loose a fit on shaft and probably badly tapered. Outer showed no motion in housing.

*(2)   Alternate: rear brg outer sticking in housing, building up reverse thrust and then letting go. Perhaps building up thrust in opposite direction.

Task Corp.   Pump-Motor 56383   Page A4

EXAMINATION OF COPPER-PLATED TEST BRGS.

These brgs all showed just what might be expected (except for larger thrust on small end brg) and nothing like what had shown on failed 204 size. In each case a fairly narrow uniform width of contact. Measurements and calculations follow: (fig. page 6, 1st report)

	With Spring alone		With pump	
Brg. No.	no. 2	no. 3	no. 1	no. 4
Size	204HJB	203S	204HJB	203S
inner: a,		.045 .058		.050 .050
b at 90°		.045 .055		" "
		.045 .058		" "
		.045 .060		" "
average	.0620 .046	.045–.0578	.060–.058	.050–.050
Outer: a, b		.030–.050		.065–.035
		.038–.050		.065–.035
		.050–.042		.057–.042
		.045–.050		.060–.038
average	.068–.035	.0408–.048	.060–.042	.0618–.0375
Load/ball				
Inner	19	40	38	30
Outer	20	26	33	12
½a + b: inner	30°	30.6	31.3	31.2
Outer	29.2	27	27.5	35.4
Contact angle				
Inner	19.5°	17.9	18.2	17.4
Outer	20.2	21.5	22	13.4
Measured	18.3	15.0	18.0	15.5
(initial by BII)				
Total Thrust				
Inner	69.5	98	131	72
Outer	76	76	128	22**

Conclusions: Thrust on main brg about where they should be: (note about 73 lb. spring alone, and 130 with pump).
Thrust on small brg always larger than expected.

** (These readings show small but positive outer tilt).

Thomas Barish   Dec. 26, 1963

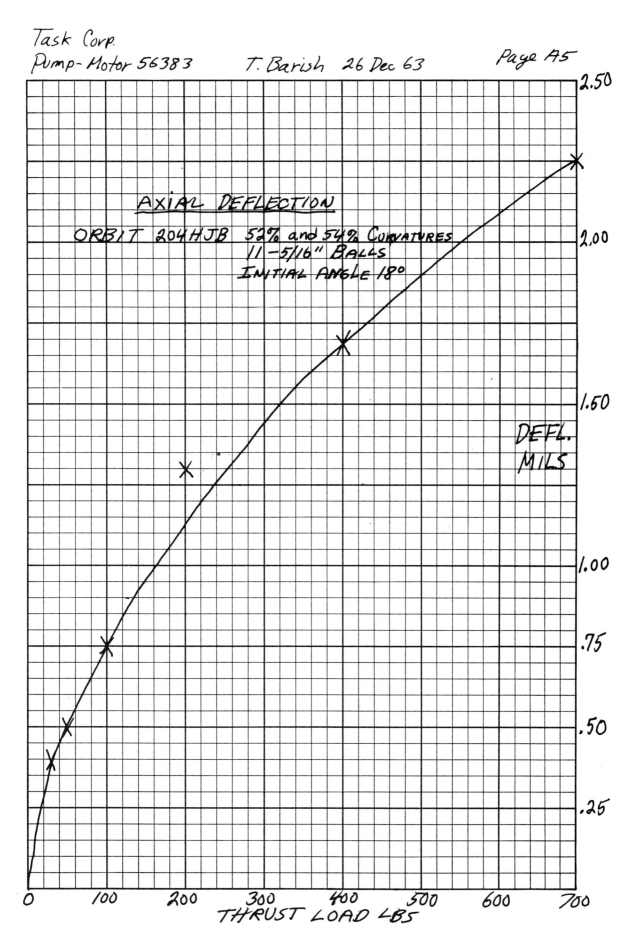

Task Corp.
Pump-Motor 56383   T. Barish   26 Dec 63   Page A5

AXIAL DEFLECTION

ORBIT 204 HJB 52% and 54% CURVATURES
11 - 5/16" BALLS
INITIAL ANGLE 18°

DEFL.
MILS

THRUST LOAD LBS

Task Corp.   Pump-Motor 56383   Page A6

OTHER POSSIBLE CAUSES:

(1)   Critical Frequencies: axial: Rotor Weight,   5.5 lb.
Spring Constant taken from curve page A5, for 100 lb. load
          = 187,000 lb. per inch.
          Deflection under own weight = 5.5/187,000 = .000 0294
          Critical Frequency = $197.7/(.000\ 0294)^{1/2}$ = $\underline{38000\ per\ min.}$**
          Forcing functions: 6000 for rpm and $\underline{54,000}$ for 9 cylinders.

(2)   Shock Effect: if rear brg hangs up and lets go quickly:
          Total Travel: estimated at .012″ max. (endplay of 204 brg.)
          Energy build up: 70 lb. spring × .012″
          Using area under deflection curve, page A5, this would require brg force
          to go from 100 lb. to $\underline{660}$ lb. max. shock
          This is still not excessive since brinell capacity or shock capacity of
          the brg is over 2000 lb.

** (Note: 3'5° Brg will increase spring rate about 60 to 70% and will raise this critical to 48,000 to 50,000 per min.)

Thomas Barish   Dec. 26, 1963

Exhibit 2 — Jack Wireman (Part 4)

for the Task Corporation, 1009 East Vermont, Anaheim, California
                    Attention: Jack Wireman, Proj. Engr.

Motor-Pump 56383: Bearing Failures.
Supplement B to Report of Dec. 14, 1963; Review of 5 additional Brgs.
Orbit 204HJB1519 that had seen small amounts of use.
(1)   *Bearing examination* on page B2. No deterioration. But ball paths just
distinguishable in 4 cases.

Conclusions:

(2)   The thrust loads were about where they should be, perhaps a little high,
but nothing to cause any trouble.
(3)   There were no signs of any tilt or misalignment of ball path.
(4)   However two of the bearings (and slightly in a third) showed that at
some time, the small end bearing took most of the thrust and left this large
end bearing under pure radial load; in one case leaning slightly towards
reverse thrust.
(5)   This confirms previous analysis (emphasized in Supplement A) that the
major cause of trouble is seizing of outer race of the small end bearing preventing axial sliding.
   And that this trouble is aggravated in longer runs by small heavily loaded
spline held in one place and tending to imbed;
   And by possible eccentricities at Pump-Motor joint causing larger radial
load.
(6)   One possible alternate for this lot; if motor was run without pump or
spring load, bearings would show radial load as above.
(7)   Previous discussion only considered as source for binding of O.D. of
small bearing, the possible greater temperature of shaft and bearing over
housing. One other possibility now suggested is that unsymmetrical end-bell
under fairly large pressure load distorts bearing housing out-of-round.
(8)   *Recommendations* still as in Supplement A:

Examination of Bearings:

5 Orbit 204 HJB1519 that had seen small amounts of miscellaneous use.
(numbered 1 to 5 by TB)

None of the bearings showed failure: practically no deterioration. In fact, they could be reassembled after very light polishing, with new (high quality) balls and used as new.

The *cages* were all in practically perfect condition. The balls all showed very light banding; one set, except 2 sets in bearing 2. The bands were very faint in bearings 4 and 5.

The *races* showed no signs of any motion on bores, O.D. or sides. However, the ball paths could be distinguished and were measured and checked for load below: (except no. 4 where marks were too faint.)

The major point of interest was that bearings 2 and 3 showed beside the standard thrust load path, a *second lighter area,* a small scraping on the inners, under approximately radial load. In no. 3, this even edged slightly towards the wrong side thrust. No. 5 also showed a very faint spot of radial load on the inner. On the outer races, these were not discernible because the polish was not so high and it would take more running to show these faint paths.

BALL PATH MEASUREMENTS AND LOAD CALCULATIONS

Bearing No.	1	2	3	5
Inner Race a	.060	.060	.060	.060
b	.050	.050	.045	.040
a + ½b	.085	.085	.0825	.080
Contact Angle, degrees	19.5	19.5	20.5	21.3
Load per Ball Lb.	22½	22½	17	22½
Total Thrust Lb.	83	83	66	90
Outer Race a	.055	.065	.065	can't see
b	.040	.035	.040	
a + ½b	.075	.0825	.085	
Contact Angle, degrees	23	20.5	19.5	
Load per ball	28	20	28	
Total Thrust, Lb.	120	78	102	

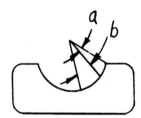

Thomas Barish
December 31, 1963

Six OUTCOME REVIEW

The main purpose of the single-chapter problem cases up to this point has been to concentrate on opportunities to practice engineering thinking and skills in situations where outcomes of the real design projects were not known. Feedback therefore had to come from self-reflection on how the engineering was performed, and from critical review by colleagues and instructor. No relief was available in the sense of being able to read ahead and learn how someone else solved the problem.

There is also value in being able to learn nature's verdict about an engineering design, and with the benefit of that verdict to consider how the engineering could have been done better, or whether it was perhaps an example of such excellence that it should serve as a precedent in the future. Physical examples of engineering design that could be considered for this purpose abound in the real world, and for the most part these are designs that have worked successfully.

Whether they work better than other design approaches that were not pursued — and if so, why — is not always easy to ascertain until a better design is found, so the knowledge of a given outcome necessarily tends to beg the question of whether it is "best" or not. Sometimes it is not even clear why the design works, just as nobody seems to know why aspirin works, though it is almost universally prescribed and used. An engineer who was designing an automobile fuel-flow meter, for example, was struggling to eliminate errors produced by formation of gasoline vapor. After some trial and error he found a modification of the design that did succeed in curing the problem. Much to his surprise he found that the solution to this problem made the meter's response to flowrate more linear, as well. He did not know why, and as he went on to the next problem he simply counted it lucky that the happy coincidence, which he concluded could just as easily have gone unfavorably, had occurred.

The last part of Jack Wireman's story of the bearing problem, which appears in this section, provides a somewhat similar outcome. By what Mr. Wireman regarded as largely good fortune the solution he tried worked successfully and became the final design, at least until he or someone else comes up with a better one. He was not sure whether it worked for the reasons he hoped it would, and if so which of them accounted for it. He also did not undertake a program of investigation to learn the answers to these questions, and perhaps it can be a matter for exploration as to whether he should have, and if not, under what modifications of his circumstances he should have.

Examples of outcomes in which causes of failure and remedies were more clearly apparent appear in the Flight Safety Foundation Design Notes of this section following the Wireman case. One of these, however, also illustrates a situation in which the sequence of failure was so destructive that the pathology could not be retraced with certainty. Ability to extract and apply useful precedent where it is apparent in outcomes, and to operate effectively in spite of uncertainty when clear precedent is not available, is thus a valuable trait in engineering work.

Exercises

1 Name some physical devices whose engineering was completed some time ago. What kinds of developments might introduce another chapter in their engineering development?

2 Describe in fantasy how a "last chapter" should look for one or more of the cases studies thus far. Discuss the factors on which you expect the nature of the engineering outcome would most depend.

3 For one or more products in your experience, describe a design weakness that appeared in use, what the potential impact of such a weakness was, and how it was remedied. (Possibly ask a new-car dealer about recalls.) What precepts would you draw from this experience?

JACK WIREMAN

Failure of a Ball Bearing

Part 5

To test the opinions advanced by the different experts as to why the bearings had failed, Mr. Wireman conducted several investigations. In his judgment, the results of these investigations cast doubt on all the opinions advanced, and verified none. Some action, however, had to be taken to solve the problem. Mr. Wireman rejected the idea of trying several different bearings at once. Such an approach seemed to him to be excessively expensive and sloppy engineering. His decision was to try another standard bearing, New Departure No. 030204, which had a higher load capacity, a contact angle of 35°, larger balls, and an inner-race-centered, plastic retainer. The thickness of the retainer was great enough so its point of contact on the balls was halfway between the inner and outer races. The 030204 was manufactured to only a standard ABEC (Annual Bearing Engineers' Committee) class-3 quality, a class compared to ABEC class 7 for the Orbit bearings, and its price consequently was around $3 as compared to around $10 for the Orbit bearings.

As of February 24, 1964, a test motor had run with the 030204 bearing for 600 hr, and no failure had yet occurred. The main question in Mr. Wireman's mind was what he would do if the bearing did fail before the required 2,500-hr test was completed.

The possibility that the rear bearing was misaligned or binding was eliminated by careful measurement of the parts concerned while operating, and computations as to the expansion effects to be expected with higher temperatures. Computations indicated that the aluminum motor housing could be expected to expand from 0.003 to 0.006 in. farther axially than the shaft as temperature rose to that of normal operation. Mr. Wireman concluded that there was more than enough room for this much axial movement of the rear bearing, and movement would be made easier at higher temperature because the aluminum housing would expand radially more than the outer race.

Careful examination revealed no evidence of spline imbedding. Axial loading of the shaft under operat-

ing conditions was checked by measuring how far the shaft moved with the motor running. Exhibit 1 — Jack Wireman (Part 5) shows schematically how this measurement was made. No excessive deflection resulted, and it was therefore concluded that the maximum load on the shaft was not excessive.

Varying the applied axial loading by varying throttling of the hydraulic-fluid outflow, the shaft was forced to move about 0.008 in. back and forth with the motor running, and it was observed that the rear bearing slid longitudinally in the housing, as it was supposed to, without binding. Observation through a Plexiglas cover substituted over the rear bearing made it possible to see that the belleville load was sufficient to keep the rear bearing balls from skidding.

The possibility of ball skidding of the front bearing was also studied. The front of the motor was replaced with a similar transparent plastic through which the bearing could be observed. The motor was then filled with oil, and run. With stroboscopic light, it was possible to tell by observing the retainer whether the balls were revolving at the right speed or going too slowly and skidding. By varying the axial load hydraulically, the loading at which skidding occurred could be measured. It was found that the 204HJB1519 bearing would not skid unless the axial load dropped below 60 or 70 lb. With the lighter load 204SST5 bearings, only 20 to 30 lb were needed to prevent skid. With the 030204, around 40 lb were needed. All these figures were less than the loading expected in operation, so Mr. Wireman concluded that ball skidding was not the cause of failure.

Mr. Wireman also looked for signs of cavitation in the hydraulic fluid, but he could not see any. He surmised that the pressure of 50 psi in the motor was enough to prevent cavitation.

These tests were performed on the Task dynamometer. It was not possible to conduct them with a pump attached to the motor, but this fact, in Mr. Wireman's opinion, did not cast doubt on the validity of the results.

Another possibility that had occurred to Mr. Wireman was that there might be some sort of high-frequency-pressure feedback from the Geyser pumps. He had asked Geyser about this, and was told that no such feedback was present. The Geyser representative commented, however: "We've been told that bearing problems have occurred with that type of pump before. The pump wails like a banshee when it's running. Something has to be vibrating quite a bit."

If the bearings were failing due to excessive loading imposed by the pump, it was presumed that Geyser Pump might be liable for the expenses of determining the cause of the bearing problem and for correcting it. Such liability would have to be determined on a negotiated basis, however, since the original specification to which the motor had been designed had made no reference to axial loadings on the motor shaft due to the pump.

As of mid-March, 1964, all the pump motors were being assembled with the New Departure 030204 bearing. A test motor was also being run in the Task shop with this bearing, and after 600 hr no problems had yet occurred. It had been decided that all the bearings should be "sound tested" by the Bearing Inspection Company before installation. At first a rejection rate of 50 percent was experienced with this inspection. The cause was found to be that the supply house from which the bearings were bought had been repackaging them with insufficient care. In repackaging, some of the bearings had apparently become exposed to moisture, which caused slight corrosion and consequent roughness of the balls and races.

Aircraft were already flying with the pumps installed, some having the Orbit 204HJB1519 and some having the New Departure unit. Maintenance instructions on the aircraft were set to require teardown and inspection of the pump motors after 1,000-

hr operation, but none had yet been in use this long. A couple of pumps had failed during preflight testing due to clogged outlet filters (10-micron filters), but none had yet suffered bearing failures. Some other problems with the motors, such as burning of the electrical connector pins, need for higher flow-rate, faster cold starting, and lower current drain had been corrected by various design modifications, none of which was expected to affect shaft-bearing life. Not all of these changes were Task's responsibility, and the costs of making them had accordingly been divided among Task, Geyser Pump, and Thunder Aircraft by mutual agreements.

Over 120 motors had now been shipped, and it had been agreed among Task, Geyser Pump, and Thunder that Task was responsible for performance of the bearings. There were 180 motors yet to be completed and shipped on the contracts.

Discussion Questions

1   What should Wireman have done next if his last solution had not worked?

2   How should the costs of delays and experimentation involved in the bearing problem be shared among Task, Thunder, Geyser, and Orbit? Explain your rationale and describe how each of these four parties might react to it.

3   What are the pros and cons of conducting further research to explain the failures? To what extent should each of the four companies be willing to finance such research?

4   What should Jack Wireman have learned from this experience? What did you learn from it? Which of you should have learned more? Explain.

Exhibit 1 — Jack Wireman (Part 5)

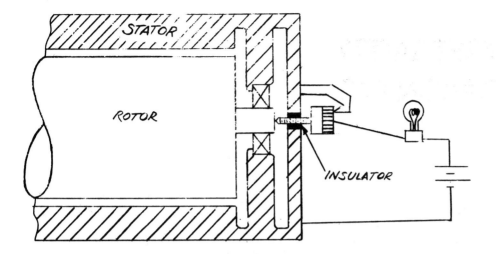

FLIGHT SAFETY FOUNDATION, INC.

Flight Safety Foundation, Inc., was formed in 1945 for the purpose of promoting aircraft safety. It is supported by contributions from U.S. industrial firms, particularly insurance companies, but also oil companies and others.

Activities of the foundation include the publishing of safety bulletins for pilots, mechanics, and others concerned with both civilian and military aircraft operations in the United States and abroad. Design notes such as those in Exhibits 1–10 — Flight Safety Foundation were intended to inform aircraft designers about the kinds of design weaknesses that have led to undesirable outcomes in aircraft operations. Each note was gathered from circumstances of a real problem. The notes deal only with the engineering outcome and what ensued, not with how engineering was performed (or by whom) leading to the design weaknesses.

Two types of notes are included here. The first five describe the "fixes" applied and precepts drawn by writers of the notes, as well as the failures that occurred and what the physical causes appeared to be. The second five notes emphasize mainly the failures and causes, without giving details of the solution or conclusions about precepts. In either of these two note types the reader may wish to draw conclusions of his own as well as considering inferences of the note-writers.*

Discussion Questions

1   Suggest how stated precepts of the first five notes apply to other cases studied in this book. Can you suggest alternate precepts for any of the first five notes?
2   For the second five notes formulate:
    a   appropriate fixes (how many can you suggest for any given note?)
    b   precepts
    c   suggestions as to how these precepts might apply to other cases in this book and to other devices you have encountered in your experience
3   One of the notes makes reference to Murphy's Law. What kinds of factors in the world cause formulation of such a statement? What should be an engineer's responsibility with regard to such factors?

* The design notes 55–2, 55–12, 56–3, 56–6, 58–8, 61–4, 64–5, 65–1, and 67–7 appear here by permission of the Flight Safety Foundation, Arlington, Va.

Exhibit 1 — Flight Safety Foundation

POWER PLANT CONTROLS —
Cowl Flap

design notes

Hole Burned in Tube Nullified Emergency Oxygen System

the Situation

A BRILLIANT FLASH was noticed in the vicinity of the flight engineer's switch panel at the same time the circuit breaker in the cowl flap motor control circuit tripped, following the use of manual operation when one engine's automatic cowl flap control system was found to be inoperative.

Lowering the switch panel revealed a hole burned in the tube connecting the captain's emergency smoke mask to the oxygen supply. Because of insufficient clearance, the metal tubing touched an unprotected relay terminal, shorted out the cowl flap circuit and burned a hole in the tubing.

the Hazard

The incident contained two hazards: failure of the cowl flap system to operate because of a short circuit, and the chance of a serious fire had the line contained oxygen at the moment. Fortunately the first hazard was easily remedied and the second did not materialize.

Note: Apparently, designers are not always aware of the extreme precautions necessary to prevent accidents involving oxygen systems and equipment. In this instance, two incompatible systems — electrical and oxygen — were installed in close proximity to each other, an oversight that systems designers should be careful to avoid.

OXYGEN LINE

RELAY

COWL FLAP CIRCUIT BREAKER

FLIGHT ENGINEER'S SWITCH PANEL

Hole burned

in oxygen tube

the Fix

The oxygen line was rerouted and properly secured in place.

PRECEPT

The structure and its components should be compatible — one part with the other from the standpoint of durability, deflections, wear, and the danger of one creating a hazard by proximity to other parts.

Ref: DESIGN NOTES, Bulletins 52-7, 53-3, and 53-8

FLIGHT SAFETY FOUNDATION, INC.
468 FOURTH AVENUE • NEW YORK 16, N. Y. • • • WESTERN OFFICE: 2038½ GRIFFITH PARK BLVD., LOS ANGELES 39, CALIF.

BULLETIN NO. 56-3

Exhibit 2 — Flight Safety Foundation

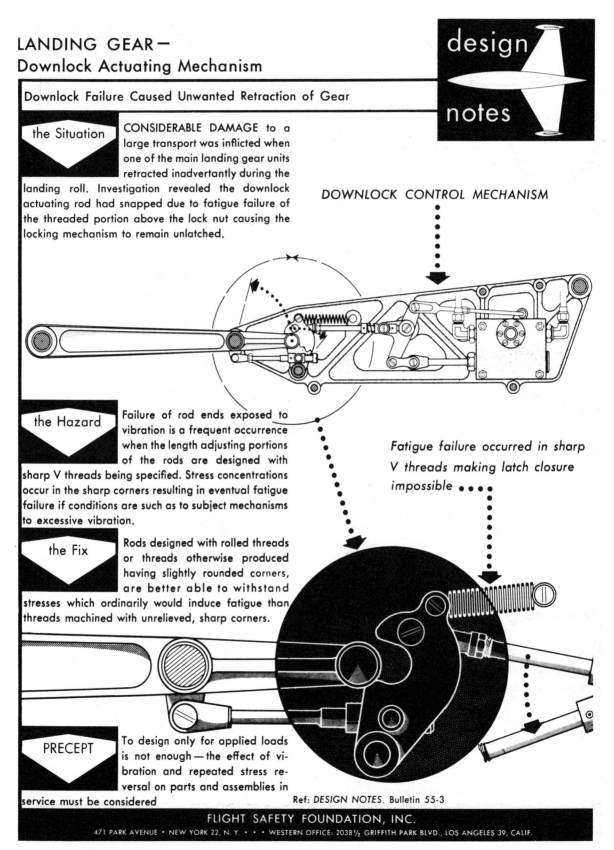

LANDING GEAR—
Downlock Actuating Mechanism

Downlock Failure Caused Unwanted Retraction of Gear

the Situation — CONSIDERABLE DAMAGE to a large transport was inflicted when one of the main landing gear units retracted inadvertantly during the landing roll. Investigation revealed the downlock actuating rod had snapped due to fatigue failure of the threaded portion above the lock nut causing the locking mechanism to remain unlatched.

DOWNLOCK CONTROL MECHANISM

the Hazard — Failure of rod ends exposed to vibration is a frequent occurrence when the length adjusting portions of the rods are designed with sharp V threads being specified. Stress concentrations occur in the sharp corners resulting in eventual fatigue failure if conditions are such as to subject mechanisms to excessive vibration.

Fatigue failure occurred in sharp V threads making latch closure impossible

the Fix — Rods designed with rolled threads or threads otherwise produced having slightly rounded corners, are better able to withstand stresses which ordinarily would induce fatigue than threads machined with unrelieved, sharp corners.

PRECEPT — To design only for applied loads is not enough — the effect of vibration and repeated stress reversal on parts and assemblies in service must be considered

Ref: *DESIGN NOTES*, Bulletin 55-3

FLIGHT SAFETY FOUNDATION, INC.
471 PARK AVENUE · NEW YORK 22, N. Y. · · · WESTERN OFFICE: 2038½ GRIFFITH PARK BLVD., LOS ANGELES 39, CALIF.

BULLETIN NO. 55-12

Exhibit 3 — Flight Safety Foundation

SURFACE CONTROLS—
Rudder and Brake Pedals

design notes

Potential Danger Lurked In Exposed Control Mechanisms

the Situation

IMPAIRED OPERATION of rudder pedals and brake controls was experienced at frequent intervals but always discovered in time to avoid any serious consequences in flight.

Investigation revealed that at various times bits of hardware had dropped into the control mechanisms from above, some lodging firmly in moving parts and only released with some difficulty. These oddments consisted of spoons, metal pencils, fountain pens, electrical plugs, bolts, and unaccountably in one instance, a lipstick holder.

the Hazard

Whenever even a remote possibility exists of control mechanisms being fouled it is likely that at some time a loose object will intrude itself into a position where it could cause a serious accident.

Mechanism vulnerable to jamming by loose objects

the Fix

The airline operator installed a flexible canvas catch-all above the rudder and brake control installation.

PRECEPT

The structure and its components, especially moving parts, should be protected against the effects of environment likely to be encountered in service such as vibration, dust, loose objects, oil mists, and temperature extremes.

CANVAS CURTAIN

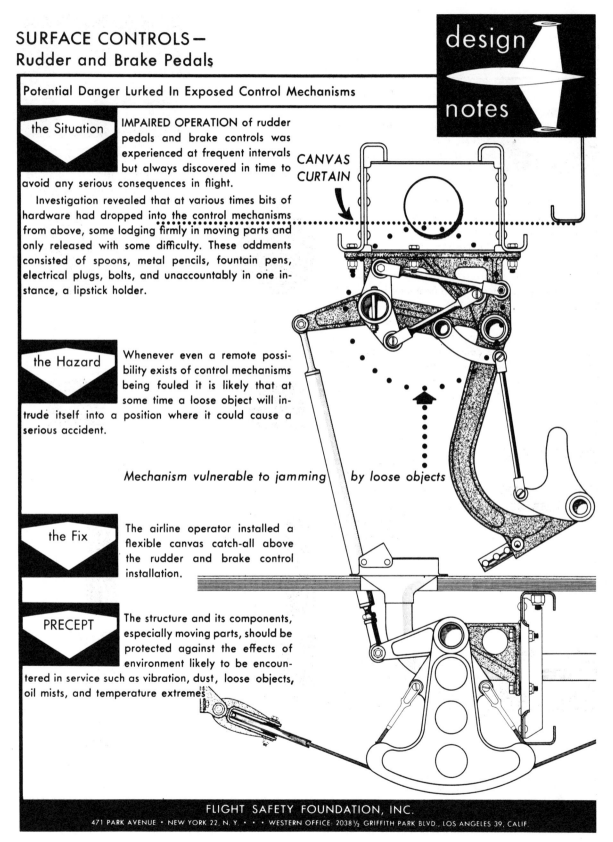

FLIGHT SAFETY FOUNDATION, INC.
471 PARK AVENUE • NEW YORK 22, N. Y. • • • WESTERN OFFICE: 2038½ GRIFFITH PARK BLVD., LOS ANGELES 39, CALIF.

BULLETIN NO. 56-6

Exhibit 4 — Flight Safety Foundation

ELECTRICAL SYSTEM—
Booster Pump Circuit

Reversed Power Leads Caused Fuel Tanks To Explode

ONE FATALITY, two seriously injured airmen, and two aircraft destroyed in explosions occurring during replacement of fuel booster pumps.

Examination of the wreckage revealed the d.c. power leads were reversed. Consequently, when the correctly wired replacement pumps were connected to the reversed leads, the positive side went directly to the ground.

The accidents occurred at the time the motor circuits were being tested prior to final installation of the pumps in the fuel tanks. The airman who was killed held the pump in the tank opening — the others had loosely bolted their pump to the mounting flange. In both cases the loose connections caused an intense arc which ignited the explosive fuel-air mixtures when the current was turned on.*

In several other pump installlations found with reversed battery connections, the error had been compensated by reversing the leads to the motor! If the pumps were later replaced and correctly installed, similar explosions could have occurred.

PRECEPT Procedures for adequate maintenance and operating practices established by the designer should be consistent with average human effort, ability and attitude.

*Ref: Directorate of Flight Safety Research, Office of The Inspector General, USAF, Norton AFB, Calif.

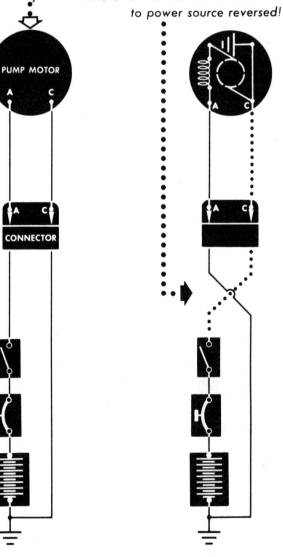

The majority of production aircraft were wired this way ••••• while a few others had connections to power source reversed!

PUMP MOTOR

Power shorted to grounded pump along dotted line

FLIGHT SAFETY FOUNDATION, INC.

468 FOURTH AVENUE • NEW YORK 16, N. Y. • • • WESTERN OFFICE: 2038½ GRIFFITH PARK BLVD., LOS ANGELES 39, CALIF.

BULLETIN NO. 58-6

Exhibit 5 — Flight Safety Foundation

SURFACE CONTROLS—
Elevator Control System

design notes

Reversed Bolt Jammed Elevator Controls

The SITUATION

Shortly after takeoff, a bomber rotated rapidly into a nose high attitude which the crew could not immediately correct. Nose down elevator pressure was applied without effect — the control column was apparently locked in the full back, nose up position. After a considerable lapse of time in which various combinations of landing flap settings, stabilizer trim tab, airspeed, air brakes, were tried, sufficient control over longitudinal attitude was available to attempt landing with reduced fuel load.

On the first attempt, the nose dropped sharply because the stabilizer trim action was not rapid enough to counteract it. Four more attempts at landing were made, the fifth being successful.

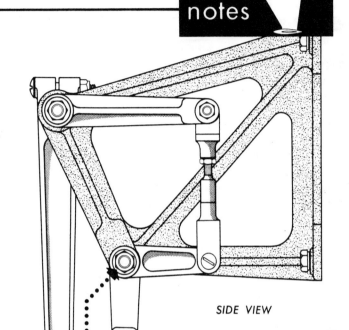

SIDE VIEW

The HAZARD

Investigation revealed the elevator control lock-up was caused by a bolt installed in reverse: the threaded portion protruded, effectively jamming a control lever.*

The design of the mechanism failed to provide against the bolt being installed in reverse.** Minimum clearance between the lever and bolt head existed when the bolt was installed correctly; reversing the bolt reduced clearance to interference.

How would you design this so that the bolt could not be installed incorrectly?

Lever moved freely when this bolt was installed in this way but, when put in reversed, jammed the lever

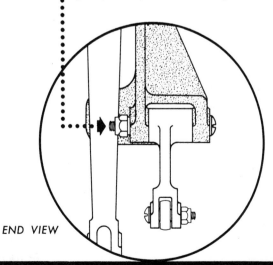

END VIEW

* Ref: Directorate of Flight Safety Research, USAF, Norton AFB, Calif.
**Murphy's Law: *"If an aircraft part can be installed incorrectly, someone will install it that way."*

FLIGHT SAFETY FOUNDATION, INC.
468 FOURTH AVENUE • NEW YORK 16, N. Y. • • • WESTERN OFFICE: 2038½ GRIFFITH PARK BLVD., LOS ANGELES 39, CALIF.

BULLETIN 58-8

Exhibit 6 — Flight Safety Foundation

INSTRUMENTS —
Circuit Disconnect Panel

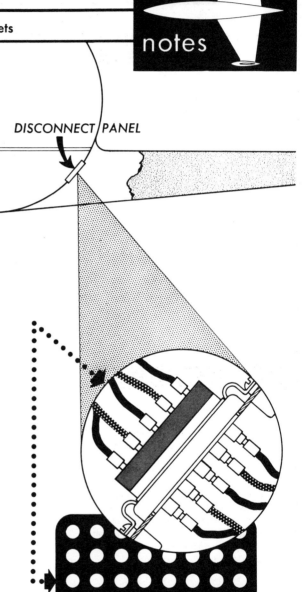

design

notes

Carbonized Rubber Pad Shorted Terminal Sockets

the Situation

CIRCUIT TROUBLES, difficult to trace, were experienced with fire warning systems, engine cylinder temperature indicators and other powerplant circuits leading to instruments in the cockpit. Inaccurate instrument readings, false fire warnings, were only part of the difficulties experienced before finally locating the cause in the electrical disconnect panels installed below the cabin floor.

These panels were mounted in a tilted position. Realizing the probability of loose parts falling on the face of the panel and shorting out the terminals, designers devised a rubber pad to fit over the sockets and thereby eliminate possibility of short circuits. Further precautions were taken to prevent moisture from collecting by cementing the pad to the panel. Despite these precautions — the result of commendable design foresight — the chemical composition of the cement defeated the purpose of the pad. Chemical reaction took place after a time, resulting in the rubber becoming sufficiently carbonized to form a high resistance short between the circuit terminals.

DISCONNECT PANEL

the Hazard

Nothing can safely be overlooked: even chemical reaction between different materials — as in this case — can present a hazard.

the Fix

A cement of a composition having no unfavorable reaction on the rubber compound of the pad was substituted for the one used originally.

PRECEPT

Aircraft structures, mechanisms and individual parts should be designed to function safely and dependably over long periods of repeated use.

Rubber composition pad cemented to panel, protected exposed sockets from being shorted by loose metal objects dropping on them.

THE DANIEL AND FLORENCE GUGGENHEIM AVIATION SAFETY CENTER AT CORNELL UNIVERSITY
471 PARK AVENUE · NEW YORK 22, N. Y. · · · WESTERN OFFICE: 2038½ GRIFFITH PARK BLVD., LOS ANGELES 39, CALIF.

BULLETIN NO. 55-2

Exhibit 7 — Flight Safety Foundation

LANDING GEAR–
Axle Design and Install. Details

The SITUATION

A LANDING GEAR AXLE BROKE just as the aircraft started the takeoff roll. The fracture, which apparently had started on the bottom of the axle at the inboard wheel bearing journal, appeared to be the result of embrittlement caused by stress corrosion. A number of similar incidents have been reported of axles snapping after the aircraft had been standing fully loaded on alert status for several days without having been moved. One case of failure occurred while the aircraft was being washed.

The HAZARD

In the majority of cases, the axles collapsed because of fatigue damage, but in others as there was no evidence of fatigue, failure was attributed to overloading. During the investigation of these incidents, grinding burns were discovered on some broken axles. According to information in the report: "Grinding Damage in High Strength Aircraft Fasteners"*, grinding burns may prove to be a serious safety hazard that heretofore has not been generally known to exist.

The report states:

A very serious and detrimental condition exists in many high strength steel bolts in use today. This is the presence of severe grinding stresses or burns. Such damage occurs when friction creates heat without proper grinding control. The increased temperature can lead to localized surface hardening or stressing.

When a metal surface comes into contact with a rotating grinding wheel, heat is generated which raises the surface to very high temperatures. Due to the speed of the operation the heat is concentrated on the very surface of the metal ... When the tempering temperature is exceeded the surface metal is softened.

Still higher temperatures, which are very common in grinding, will cause the metal to be heated above its hardening temperature ... When this occurs, the rapid heat transfer within the metal itself will cause the surface layers to transform into a hard untempered steel. These hard spots are very brittle and prone to fatigue failures as well as failure by hydrogen embrittlement.

As mentioned previously, the high heating in grinding is localized and limited to the metal surface. The outer fibers which are in contact with the wheel want to expand when heated, while the adjacent areas resist this movement since they remain at a much lower temperature . . . .

*Ref. STANDARD PRESSED STEEL CO.
Jenkintown, Pa. and Santa Ana, Calif.

The resultant stress patterns tend to put the surface of a ground metal part under tension forces. These are detrimental stresses which want to fracture or pull the metal apart . . . They subtract from the overall load-carrying capacity of the part. For example, a part that is heat treated to 160,000 psi and has residual tension stresses of 100,000 psi needs only 60,000 psi from external loading to exceed the tensile strength of the surface fibers.

COMMENT

Components such as landing gear axles that are subjected to centerless grinding which of necessity occurs as a final operation following heat treatment, may end up with a greatly reduced design margin of safety unless adequate precautions are taken.

••••

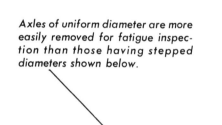

Axles of uniform diameter are more easily removed for fatigue inspection than those having stepped diameters shown below.

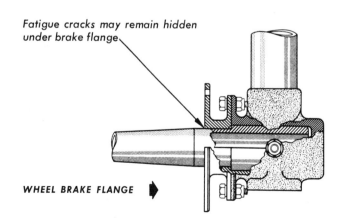

Fatigue cracks may remain hidden under brake flange.

WHEEL BRAKE FLANGE ▶

SECTIONAL VIEW OF STEPPED DIA. AXLE ASSEM.

FLIGHT SAFETY FOUNDATION, INC.
468 PARK AVENUE SOUTH • NEW YORK 16, N. Y. • • • WESTERN OFFICE: 2038½ GRIFFITH PARK BLVD., LOS ANGELES 39, CALIF.

DESIGN NOTE 67-7

Exhibit 8 — Flight Safety Foundation

ELECTRICAL SYSTEM—
Aircraft/Missile Ground Support Power

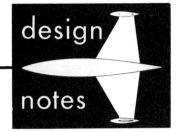

The SITUATION

SHORTLY AFTER POWER WAS CONNECTED to start the engines, the crew noticed an odor of burning insulation in the flight deck. Ground equipment power was disconnected and the engines shut down. Smoke coming from a large bundle of electrical wiring revealed the source of trouble. Several wires surrounding a power lead in the closely-packed bundle were severely burned. Evidently, the negative power lead of the 112-volt circuit had become overheated during the time the engines were being started.

The HAZARD

A BREAK IN THE 28-VOLT GROUND RETURN circuit caused the trouble. For some unexplained reason, the ground lead had become disconnected from the aircraft common ground. Normally, both the 28-volt and the 112-volt power circuits are grounded at the same place. With one of the connections open, the only return path to the ground support generator was by way of the 112-volt negative lead. Since this wire has only one-third the capacity of the 28-volt lead, it was unable to carry the high current overload from the 28-volt equipment circuits without becoming dangerously overheated.

Note: The above incident is similar to one described in a recent DESIGN NOTE in which the 28-volt power circuit polarity had been reversed by the ground power plug being inserted in the power takeoff receptacle upside down.

Note inclusion of power leads with vital electrical control circuits

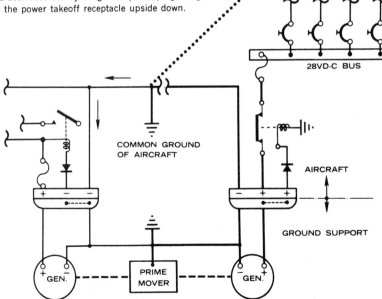

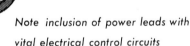

FLIGHT SAFETY FOUNDATION, INC.

468 PARK AVENUE SOUTH · NEW YORK 16, N. Y. · · · WESTERN OFFICE, 2038½ GRIFFITH PARK BLVD., LOS ANGELES 39, CALIF.

DESIGN NOTE 65-1

Exhibit 9 — Flight Safety Foundation

SURFACE CONTROLS –
Elevator Control System

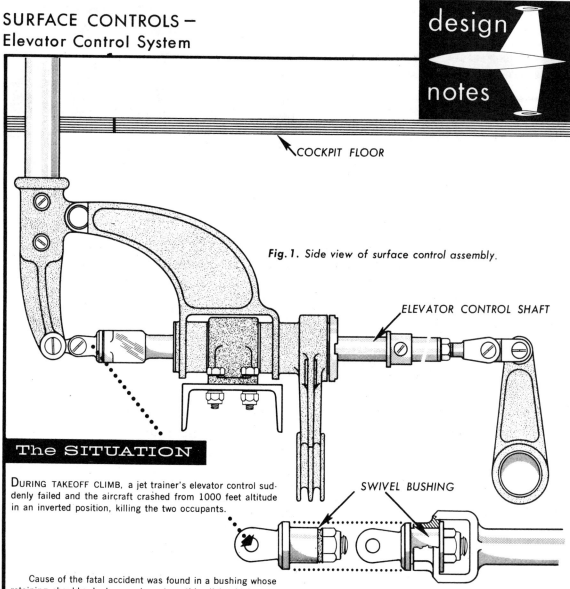

design
notes

COCKPIT FLOOR

Fig. 1. Side view of surface control assembly.

ELEVATOR CONTROL SHAFT

SWIVEL BUSHING

The SITUATION

DURING TAKEOFF CLIMB, a jet trainer's elevator control suddenly failed and the aircraft crashed from 1000 feet altitude in an inverted position, killing the two occupants.

Cause of the fatal accident was found in a bushing whose retaining shoulder had worn down to a thin disk which gave way when the control stick was actuated. When this occurred, the push rod connection separated resulting in complete loss of elevator control.*

Fig. 2. Linkage separated when bushing shoulder failed. Shaft is shown rotated from position in side view in Fig. 1.

The HAZARD

The bushing upon which so much depended was partially hidden and was therefore extremely difficult to inspect. It could gradually wear without its progressive weakening being noticed until it failed ultimately.

*Ref: DIRECTORATE OF FLIGHT AND MISSILE SAFETY RESEARCH, OFFICE OF THE INSPECTOR GENERAL, USAF, NORTON AFB, CALIFORNIA.

COMMENT

Sometimes design errors are unknowingly repeated as the above case illustrates. For instance, a comparison of the control system details of a fighter aircraft designed in 1940 by another airframe manufacturer, revealed a striking similarity of design, including the use of a bushing in a location identical to the one above.

FLIGHT SAFETY FOUNDATION, INC.
468 PARK AVENUE SOUTH • NEW YORK 16, N. Y. • • • WESTERN OFFICE: 2038½ GRIFFITH PARK BLVD., LOS ANGELES 39, CALIF.

BULLETIN 61-4

Exhibit 10 — Flight Safety Foundation

FUEL SYSTEM —
Fuel Valve Installation

The SITUATION

SHORTLY AFTER COMMENCING APPROACH DESCENT, flameout occurred followed by the fighter aircraft crash landing short of the touchdown area. The aircraft was severely damaged by both the impact and subsequent fire. The pilot escaped from the wreckage with only minor injuries.*

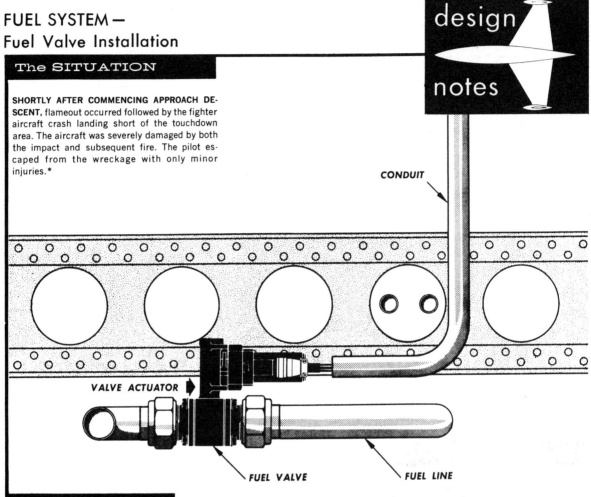

The HAZARD

THE EXACT CAUSE OF THE FLAMEOUT was never determined as most of the aircraft equipment was destroyed in the fire. Previous accidents resembling this one were known to have been caused by certain types of electro-mechanical valves accidentally shutting off the fuel. Evidently these valves were not' designed to "fail safe," consequently, if trouble developed in the control circuit wiring, or, as it frequently did, in the actuator motor, the valves would of their own accord move to the OFF position and cut off the fuel.

In this case the valve was damaged to such an extent that it was impossible to determine if it had malfunctioned, leaving only the control circuit to be examined for a possible clue. Removing the wiring contained in an aluminum conduit disclosed signs of, moisture having been trapped in a bend in the tubing close to where the wiring terminated in a connec-

tor to the valve actuator. The wiring insulation had deteriorated and inside the tubing were definite signs of corrosion.

Based on past experience, investigators assumed water had accumulated in the conduit and, following the wires, leaked into the actuator connector and shorted out circuit pins causing the valve to close. Shorting the pins would have the same effect as if the pilot had turned the fuel switch to the OFF position.

The FIX

DRAIN HOLES WERE DRILLED in conduits of existing aircraft to prevent condensed moisture from accumulating.

*Ref: DIRECTORATE OF AEROSPACE SAFETY, USAF HQ., NORTON AFB, CALIFORNIA

FLIGHT SAFETY FOUNDATION, INC.
468 PARK AVENUE SOUTH • NEW YORK 16, N.Y. • • • WESTERN OFFICE 2038½, GRIFFITH PARK BLVD., LOS ANGELES 39, CALIF.

DESIGN NOTE 63-5

Seven SERIAL-HISTORY CASES

In this section cases are similar in format to the Jack Wireman case, which has run through the book to this point, except that their parts are grouped together rather than dispersed among various topics and cases. Typically, each serial-history case part ends with one or more unsolved problems, and the next part tells what ensued. Discussion questions are suggested at the end of each case part. There is an option to struggle with the problem of each case part before going on, or simply to read all the parts first and then explore lines of inquiry. Even though you may not wish to apply major effort to the problems of one case part before moving on to the next, you are encouraged (1) to consider what questions should be pursued to cope with problems of the case situation; and (2) to think at least briefly about how you might answer or go about developing answers for questions raised both in the case and at the end of each case. This procedure should make the reading more profitable, as well as more interesting.

The three cases in this section differ from each other in several respects. The Jack Bristor case mainly poses decisional, as opposed to physical, design problems, although a creatively inclined reader may see some of the latter as well. Warren Deutsch faced a creative physical design problem for which knowledge of available types of switches had value but which could be solved with relatively little sophistication. Keith McFarland faced a diagnostic and prescriptive problem, which involved highly sophisticated mathematical analysis. Few students are equipped for this sort of analysis as undergraduates, and they are not expected to apply the techniques of Mr. McFarland, though they may find it of interest to try working through some parts of the analysis. The main intent is to give a specimen of specialized and high-powered techniques for perspective on the variety of methods used by various engineers.

Like the cases in the preceding outcome-review chapter, the serial histories offer opportunities to Monday-morning quarterback procedures of engineers to find possible general lessons and precepts.

Exercise questions in this section also are structured to provoke consideration of the profession of engineering in the perspective of particular instances.

Exercises

1 Based on their performance as you observed it in the cases, assign grades to the engineers you have seen in this book. Explain the rationale by which you assigned the grades.

2 Formulate a list of lessons learned from studying the cases. Indicate which of the lessons you learned from the case engineers and which you learned from your own attempts at analyzing the cases from the feedback of retrospect, other students, and the instructor.

3 Comment on differences between the professional engineer and a layman who likes to tinker and build things.

4 Why do MBA (Master of Business Administration) graduates with engineering undergraduate degrees typically command substantially higher starting salaries than those with other undergraduate degrees?

5 Describe the most effective curriculum you can imagine for training an engineer to assume the role of the engineer in each of *two* cases you have studied. Discuss how it would be similar and different for each of the two cases, and how compromises could be struck to serve both.

6 Describe the ideal qualifications someone would need to be the boss of one or more engineers in the cases you studied. Tell how the engineer in that case could go about obtaining those qualifications.

7 Based upon the cases you have studied, describe a generalized process for coping with engineering problems. Discuss the extent to which it is similar or different from the process for solving nonengineering problems.

8   Discuss the extent to which a computer could be programmed to perform the job of one or more engineers you have met in the cases.

9   Discuss the roles that problem definition, conceptual design, technical analysis, decision making, and outcome review can or should play in effective engineering. Illustrate with some examples from one or more cases examined thus far.

10 For two of the serial histories trace the key decisions made by the engineer and estimate roughly their spacing over time. What factors tend to determine the spacing of such decisions, and what sorts of things might the engineer be doing in between?

11 Several different ways of classifying cases or engineering experiences have been suggested in the introduction to this book and in the introduction to this section on serial cases. Identify these ordering schemes and suggest one or more others that might be practical as a way of classifying engineering experiences and learning from them. Discuss how learning from experience in a career or in research might help you to learn more about the nature of engineering processes.

COLONEL JACK BRISTOR Part 1

Open-channel Flow

River stoppages, apparently caused by ice, threaten communities near Detroit. What should the head of the district's Corps of Engineers, Jack Bristor, do?

In the winter of 1953 a sudden partial stoppage of flow in the St. Clair and Detroit rivers, which connect Lakes Huron, St. Clair, and Erie, threatened to flood several communities along their banks. The Detroit district U.S. Army Corps of Engineers was responsible for constructing and maintaining navigation channels in the two rivers, and also for doing all in its power to prevent floods along their banks. In charge of the district office was Colonel John D. (Jack) Bristor, the district engineer.

Jack Bristor was sitting in a commuter train winding its way along the Detroit River into the Brush Street Station. As he looked up from the "Detroit Free Press," he noticed that the river appeared to be frozen solid. The ice was not a single sheet, but was composed of large chunks, some as large as 6 by 20 ft and about 6 to 8-in. thick. The pieces of ice were piled in random fashion 4 or 5-ft high, and were locked together by a new layer of ice.

"I remember thinking that the rumrunners who used the ice as a highway to carry their liquor across from Canada during Prohibition would have had a tough time on that surface," he recalled. "I supposed it had been caused by the unusual pattern of winter weather. December, January, and early February were so moderate that Detroiters talked jokingly about being in the Banana Belt. Then in the middle of February it turned very cold, with temperatures down to $-10°$. There was a short thaw, followed by another cold wave."

Upon arrival at his office, Jack was greeted by Joe Darcy, chief of operations. Joe was a graduate civil engineer who had worked his way up from a surveyman on the rivers and the Great Lakes, and who had about 22 yr with the district. One of the first things Joe said was, "Jack, it looks like something is blocking the flow in both the St. Clair and Detroit

rivers. The Algonac gage has been rising since yesterday noon. [See maps in Exhibit 1 — Colonel Jack Bristor (Part 1).] Gibraltar and Ft. Wayne have been erratic, but the trend seems to be up." Jack asked about Lake St. Clair and was told that the Grosse Point Yacht Club and Windmill Point gages seemed steady, but that some of the readings were missing.

Jack then asked, "What can we do about it?"

"Gosh, I don't know. There doesn't seem to be anything we can do," Joe replied.

Bristor asked Darcy to arrange to have hourly gage readings along the two rivers and Lake St. Clair telephoned into the office. Joe said he had been to four points on the Detroit River the day before, and described exactly what Bristor had seen from the train.

"It really puzzles me that both rivers are rising," Jack said. "I wonder if aerial reconnaissance would tell us anything?"

"From what I saw yesterday, I don't think so, but it certainly can't hurt," Joe replied.

Background

Bristor had received a master's degree from Cornell about 15 yr before, majoring in hydraulics. However, his thesis related to beach erosion. He recalled having studied open-channel flow, but did not specifically remember the effect of ice, as the emphasis was on transportation and sedimentation of soil in rivers of alluvial valleys. At Cornell he had streamgaged a small river from an aerial tramway, and he had visited a catamaran survey party taking flow measurements in the St. Clair River 3 yr before.

The Army Engineers had two districts in Detroit, the Detroit district — of which Col. Bristor was head — performed construction, operation, and maintenance in the area; the U.S. Lake Survey, which had a different district engineer (but which Bristor had headed in 1949–1950), surveyed the Great Lakes and their connecting channels, studied the hydrology of the Great Lakes, and kept records of lake levels and river flows. From his Lake Survey experience, Bristor had obtained a number of facts and impressions about the two rivers and Lake St. Clair.

The two rivers varied in width from about 0.5 to 3.5 miles; through each was a navigation channel 25 ft in depth and a minimum width of 600 ft. Lake St. Clair was about 20 miles in diameter, and generally fairly shallow, with maximum natural depths up to 21 ft. A navigation channel of the same minimum size as the rivers crossed the lake. Profiles of the rivers at two points are shown in Exhibit 2 — Colonel Jack Bristor (Part 1).

Flow in the two-river, one-lake system depended on the relative heights above sea level of Lake Huron and Lake Erie, and was generally an exponential function of the difference in elevation of the two major lakes. The equations for these flows, contained in Exhibit 3 — Colonel Jack Bristor (Part 1), had been carefully checked by stream-gaging. At the moment, flows in the two rivers were at a record high for February, though they had been higher in other months. Exhibit 4 — Colonel Jack Bristor (Part 1) shows historical flowrates over a 17-yr period.

There were three gages along each river and one on Lake St. Clair. Maps showing the rivers and gage locations appear in Exhibit 1 — Colonel Jack Bristor (Part 1). The gages were spring-powered and had automatic recording charts that plotted changes in water level throughout the day, but they could also be read manually at any time. A gage-reader normally tended them once a day to wind the gages, replace the charts, and see that they were operating normally. High-precision measurements were made to calibrate the gages, and these had shown the gages to be accurate to within 0.01 ft. Each gage was located in a shelter that protected it from weather and waves. Exhibit 5 — Colonel Jack Bristor (Part 1) depicts the recording mechanism, and Exhibit 6 — Colonel Jack Bristor (Part 1) shows a typical recording taken from a gage.

Complete records of lake levels had been maintained since 1860. Some fragmentary records dating back to 1836 were also available. Consequently, by examining hydrographs (time versus water-level curves), it was possible to determine the levels of all lakes at any time within the past 92 yr. By correlating this information, gage heights at any point as well as flowrates could be determined for any time during this period.

The St. Clair River–Lake St. Clair–Detroit River system contained no regulating works (gates). There were, however, gates (and power plants) on the St. Mary's River [see Exhibit 1 — Colonel Jack Bristor (Part 1)], which connected Lake Superior and Lake Huron. These gates could vary the flow into Lake Michigan-Huron from Lake Superior from 2,000 to 130,000 cfs.

At times in the past, strong northerly winds had tended to pile up the water in Lake Huron so as to raise water elevations at the south end. This rise in "head" produced a flood crest that could be traced as it passed down the St. Clair River. Lake St. Clair damped out the crest, however, and there appeared to be no logical explanation for the current rise in the Detroit River.

Along the St. Clair River, on both the Michigan and Ontario sides, were a number of small, rather low-lying towns and cities. There was industrial and residential development on Lake St. Clair, in addition to an Air Force base. Both banks of the Detroit River were highly industrialized from Lake St. Clair downstream for a distance of about 30 miles. Beyond that, the land was used for residential and agricultural purposes. Railroad tracks from Toledo to Detroit paralleled the Detroit River, being quite close to the water in some locations. Col. Bristor realized that continued rise of water levels would probably cause severe property damage and possibly even loss of life.

Initiating Action

After his conversation with Joe Darcy, Jack Bristor telephoned his friend Bill Sticks, who commanded Selfridge Air Force Base, about 30 miles from downtown Detroit. He described what he knew of the situation and asked if Sticks would send a reconnaissance aircraft over the rivers and Lake St. Clair. Sticks, after making it clear that his aircraft were *not* under the operational control of the Army, readily agreed to have the mission flown. He offered the services of a trained aerial observer or, alternatively, to allow one of Bristor's engineers to come along. Bristor weighed the fact that the mission could be expected to get off the ground 2 hr sooner with an Air Force observer against his belief that one of his own engineers might know better what to look for. "OK, Bill, send your own man, and ask him to look particularly for any large ice jam. Probably the lower he can fly, the better. If you'd have him phone me as soon as he lands, I'd appreciate it."

At this point, the mayor of Algonac, a small city on the west bank of the St. Clair River, telephoned and said that downtown Algonac was under a foot of water, that the Army Engineers were responsible for preventing that sort of thing, and that they should do something about it right now. Bristor expressed his regret about the flooding, told the mayor that excessive rainfall over the upper Great Lakes was responsible, and gave his assurance that the matter was being investigated at the moment. [Exhibit 7

—Colonel Jack Bristor (Part 1) gives long-term gage readings at Algonac.]

"My usual approach to a difficult problem like this is to mobilize all the resources of the district, especially its best brainpower, to tackle the job," Jack commented. "I happen to believe that an engineering manager should make decisions on the basis of the fullest information and the best technical knowledge he can get. I use personal observation, study, and 'picking the brains' of my staff until I figure out what to do. Sometimes this makes me a little unpopular with some of the experts who think their word should be accepted as gospel, but I think it works."

For this reason Jack decided to call a conference of the people whom he felt knew most about lake levels and flow in the system. He telephoned each of the following, briefly describing the situation, asking him to meet in about an hour, and requesting that each bring along any useful data he could assemble in that short period:

Joe Darcy, chief of operations, Detroit district, previously mentioned.

Mel Jones, chief of engineering of the Detroit district. Mr. Jones, a graduate electrical engineer, had been with the district since the early 1920s.

Phil Ash, chief of design of the Detroit district, a graduate engineer with about 15 yr of experience, including a great deal of field work on the rivers.

Frank Brown, chief hydrologist of the U.S. Lake Survey, a graduate engineer who had devoted over 25 yr to the study of lake levels and flow in the connecting channels. Jack Bristor admired Mr. Brown's technical competence and his ability to back up any statement he made with records.

Ed Rains, assistant chief of the U.S. Lake Survey, a graduate engineer who had been a hydraulics major at Michigan State and who had a background of 20 yr of experience with matters relating to the Great Lakes and their connecting rivers.

Not present, but available by telephone, was Samuel Sloan, a graduate engineer who had come to the Lake Survey in 1902 and who had recently retired at age 70. Mr. Sloan had devoted the past 30 yr solely to study of lake levels and flow in connecting channels. He was recognized as an authority on the subject, and had recently been called back from retirement to present a paper at a joint meeting of the American Society of Civil Engineers and the Engineering Society of Canada.

At this point the chief engineer of the Detroit Southern Railroad called and said that water along the Detroit River was getting close to its tracks. The mayor of Algonac called again to say that people were traveling down the streets in rowboats. More calls were coming in. Bristor asked the district executive to field the calls while he prepared for the meeting. "I think staff conferences without some kind of agenda are a complete waste of time," he commented. "This time especially we can't afford mistakes."

Discussion Questions

1 Define the problem facing Jack Bristor in at least three different ways. How do you think he has defined it?

2 Describe the purpose of Bristor's meeting.

3 Tell how you would suggest the meeting should be conducted, how long it should take, at what point it should stop, what should happen next, and why.

Exhibit 1 — Colonel Jack Bristor (Part 1)

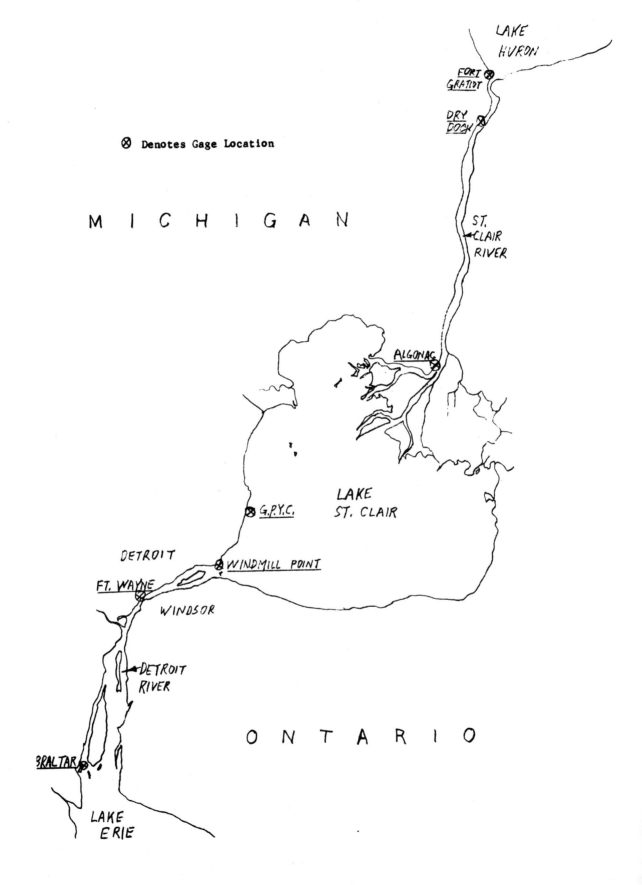

⊗ Denotes Gage Location

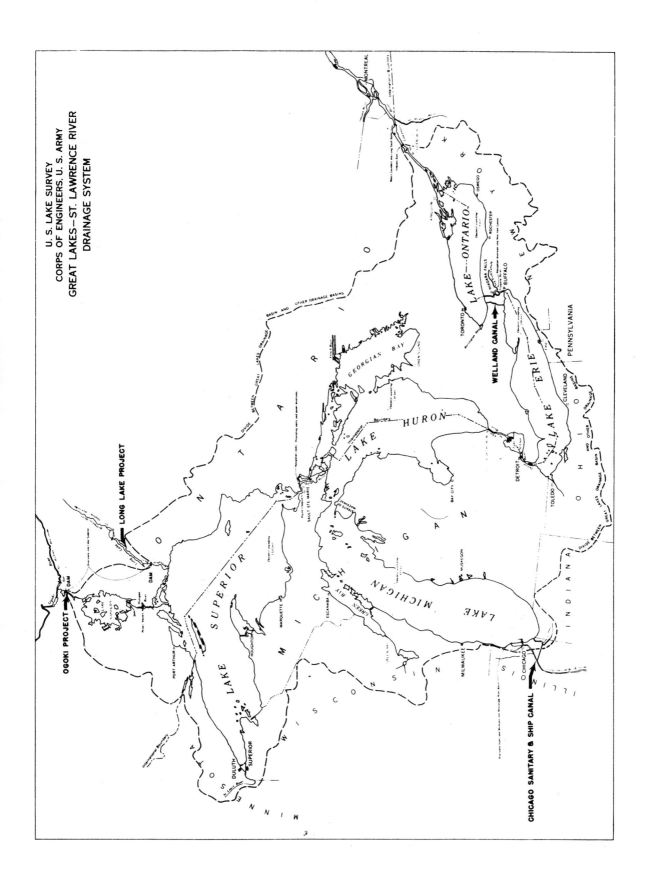

U. S. LAKE SURVEY
CORPS OF ENGINEERS, U. S. ARMY
GREAT LAKES—ST. LAWRENCE RIVER
DRAINAGE SYSTEM

Exhibit 2 — Colonel Jack Bristor (Part 1)

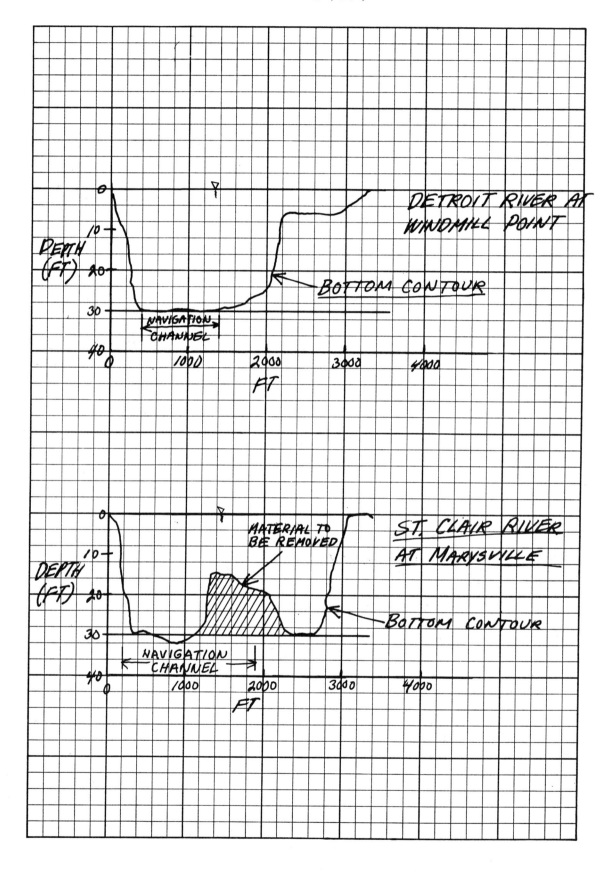

Exhibit 3 — Colonel Jack Bristor (Part 1)   Stage–fall–discharge equations for the St. Clair and Detroit Rivers

ST. CLAIR RIVER	
Period	*Equation*
1937–1958	$Q = 73.515(.5HB + .5GPYC - 541.52)^2(HB - GPYC)^{.5}$
1962–1967	$Q = 73.515(.5HB + .5GPYC - 540.84)^2(HB - GPYC)^{.5}$

DETROIT RIVER	
Period	*Equation*
1937–1958	$Q = 220.625(WP - 548.91)^2(WP - GIB)^{.4}$
1962–1967	$Q = 220.625(WP - 548.46)^2(WP - GIB)^{.4}$

HB = Harbor Beach Gage
GPYC = Grosse Pte. Yacht Club Gage
WP = Windmill Point Gage
GIB = Gibraltar Gage

Exhibit 4 — Colonel Jack Bristor (Part 1)   Historical flowrates

Corps of Engineers
U.S. Army

U.S. Lake Survey
Detroit, Michigan

Monthly and Annual Flow in Thousands of Cubic Feet per Second of the St. Clair and Detroit Rivers

YEAR	JAN	FEB	MAR	APR	MAY	JUN	JUL	AUG	SEP	OCT	NOV	DEC	MEAN
1940	124	126	141	164	168	170	172	175	176	173	173	169	161
1941	133	115	149	163	175	175	172	170	169	174	174	175	162
1942	165	99	155	182	182	188	189	186	186	184	180	177	173
1943	118	133	172	185	181	194	210	213	209	207	204	196	185
1944	145	162	169	195	197	197	201	200	199	200	194	182	187
1945	150	161	186	188	195	200	202	200	200	198	195	177	188
1946	170	143	192	201	199	199	200	200	195	191	191	188	181
1947	154	143	182	188	190	194	202	204	202	200	198	195	188
1948	168	164	189	196	202	198	198	198	192	182	177	180	187
1949	181	177	150	179	179	179	180	180	176	172	169	166	174
1950	162	138	149	171	173	176	185	187	189	189	187	185	174
1951	169	170	185	194	201	205	211	213	213	216	218	204	200
1952	200	205	208	218	221	225	228	231	230	223	216	216	218
1953	213	205	207	212	213	218	222	223	220	215	209	204	213
1954	170	154	202	203	207	212	218	217	215	222	217	214	204
1955	201	188	202	206	208	212	210	208	199	193	189	186	200
1956	122	120	170	188	194	194	196	198	194	190	187	183	178

Exhibit 5 — Colonel Jack Bristor (Part 1) Automatic water level recorder used by U.S. Lake Survey at nearly 50 gaging sites

This Cable Attaches to Float

Exhibit 6 — Colonel Jack Bristor (Part 1)

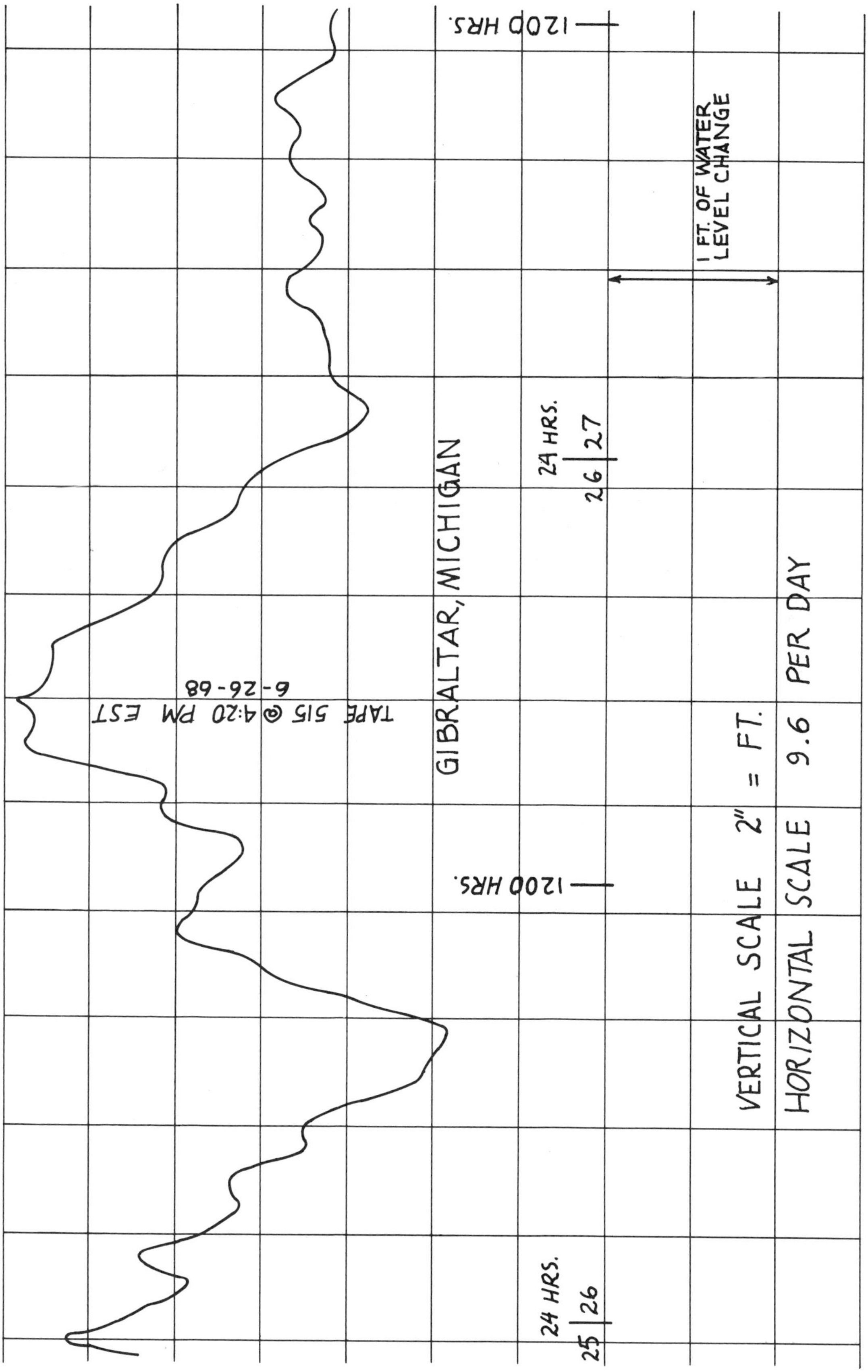

GIBRALTAR, MICHIGAN

1200 HRS.

1200 HRS.

TAPE 515 @ 4:20 PM EST
6-26-68

24 HRS.
26 | 27

24 HRS.
25 | 26

1 FT. OF WATER
LEVEL CHANGE

VERTICAL SCALE 2" = FT.

HORIZONTAL SCALE 9.6 PER DAY

Exhibit 7 — Colonel Jack Bristor (Part 1)

CORPS OF ENGINEERS, U.S. ARMY
U.S. LAKE SURVEY
MONTHLY AND ANNUAL MEAN ELEVATIONS OF
ST. CLAIR RIVER
AT
ALGONAC, MICHIGAN

YEAR	JAN.	FEB.	MAR.	APR.	MAY	JUNE	JULY	AUG.	SEPT.	OCT.	NOV.	DEC.	MEAN
1898													
1899						576.33							
1900													
1901					575.95								
1926							574.84	574.96	574.97	574.01	574.91	575.16	
1927				574.95	575.26	575.55	575.76	575.70	575.44	575.30	574.99	575.24	
28	575.27	574.67	574.42	575.41	575.61	575.95	576.35	576.37	576.20	576.06	576.13	576.05	
29		576.84	576.64	577.01	577.58	577.62	577.79	577.65	577.33	576.95			
1947											576.57	576.31	
1952								578.12	577.92	577.42	576.96	576.85	
1953	576.85	576.93	576.93	577.06	577.26	577.55	577.69	577.65	577.38	577.01	576.65	576.46	577.12
1954	576.23	576.06	576.65	576.85	576.98	577.13	577.27	577.20	577.03	577.30	577.15	576.99	576.90
1955	577.30	576.99	577.15	577.14	577.24		577.22	577.07	576.67	576.42	576.14	575.94	
1956	574.47	574.39	575.53	575.85	576.58	576.55	576.69	576.74	576.55	576.09	575.75	575.55	575.90
1957	575.29	575.33	575.49	575.63	575.86	575.99	576.34	576.15	575.95	575.61	575.36	575.36	575.70
1958	574.88	574.43	575.04	575.03	575.05	575.19	575.39	575.38	575.26	574.94	574.66	574.78	575.00
1959	574.25	574.62	574.83	575.05	575.30	575.42	575.40	575.34	575.21	575.12	575.00	575.08	575.05
1960	575.53	574.94	575.38	575.74	576.10	576.56*	576.75	576.83	576.65	576.34	575.92	575.83	576.05
1961	575.78	575.45	575.51	575.79	576.20								

* PARTIAL RECORD

FILE EED 2-1

ELEVATIONS ARE IN FEET AND REFER TO MEAN TIDE AT NEW YORK, 1935 DATUM

COLONEL JACK BRISTOR Part 2

Open-channel Flow

Jack Bristor opened the meeting by asking if the water had actually risen and if there really was a problem. Mr. Brown rejoined that there certainly was a problem. He said that tried-and-true gages, some of which had operated without repair for over 75 yr and which were being read both automatically and manually, simply could not be in error. Mr. Darcy added that the people in Algonac, which was very low in elevation, often became excited, but that the Detroit Southern Railroad engineer was a "solid citizen who would never call unless he was really concerned." Jack commented that the whole situation seemed unreal. Mr. Rains answered that the Lake Survey had studied water levels for over 100 yr, and that he was sure the present situation was unprecedented.

Mr. Jones asked for the latest information on the water levels. Mr. Brown said that the Gibraltar and Ft. Wayne gages were rising, the Algonac gage was holding steady, but that the Lake St. Clair gage at the Grosse Point Yacht Club was out of action. Bristor said he wanted to know why the water was rising, and that he believed the key to this related to the flow or the ice, or both. Mr. Brown said that the present flow, which was based on the present differential elevation of Lakes Huron and Erie, was high, but had been higher in the past without flooding. It was higher than usual for February, however.

Jack asked about the effect of surface ice on the flow of large rivers such as the Detroit and St. Clair. Mr. Rains said that not too much was known about the effects of ice, but that both rivers had been carefully measured during severe winters by stream-gaging through the ice, and that ice seemed to cause little change in the rate of flow. Actually, most people believed that the ice merely floated on the surface in the same way a boat did. Mr. Darcy and Mr. Ash said that it simply had not been cold enough for these deep rivers to have frozen solid, that actually the present basic ice cover was only 4 or 5-in. thick, and that the irregular broken pieces of ice resulting from the thaw prior to the most recent freeze were only about 6 or 8-in. thick. Bristor then asked if the windrowing of this broken ice could have formed an ice dam (wholly or partially) that could impede flow. Mr. Brown, Mr. Ash, and Mr. Rains said they were ex-river rats who had worked around the rivers winter and summer, and they had never seen it happen that way. They had talked to pilots and crews of the Coast Guard icebreakers, who said that in these large, deep, wide rivers, ice jams under conditions like the present just never seemed to come about. The entire staff agreed fully with the opinion that the surface ice was floating on top of the water, and not impeding flow.

Bristor's secretary walked into the meeting room and said, "Colonel, you just have to take this call." Jack left the room and talked to a lady who lived near the corner of Eight Mile Road and Woodward Avenue (the *highest* elevation in Detroit). She said that the streets were full of water and asked what should she do about it. Stifling an impulse to tell her to run for the woods, Bristor said he would look into it. He then asked his secretary to telephone the Detroit Police. He returned to the conference room, not without a glance outside to see if his own building were awash.

Jack quickly recalled an experience that had occurred 3 months earlier on the St. Mary's River, which he thought might have some bearing on the present situation. Three power plants and control gates regulated the flow in the St. Mary's River. The river was flowing at its maximum rate when the temperature dropped to $-27°F$. The power plants went out of operation, apparently due to a phenomenon known as frazil ice, defined as "ice crystals or granules sometimes resembling slush that are found in turbulent water." Such ice is seldom seen because it forms beneath the surface, in such places as hydroelectric inlets. Its consistency is similar to that of shaved ice, and reportedly it is very unstable. Bristor had heard it referred to as "an isotope with a half-life of 10^{-10} sec." When the power plants on the St. Mary's went out of operation, apparently the frazil ice disappeared, allowing flow to resume. The control gates were then closed partially, further reducing the flow so the power plants could be placed back in operation. With the reduced flow, conditions returned to normal.

It occurred to Jack that frazil ice might be causing restriction of flow in the Detroit and St. Clair rivers, and upon returning to the meeting he suggested this idea to the group. Mr. Jones said it was highly unlikely, frazil ice being so unstable. He could not imagine frazil ice lasting 10 or 11 hr. Further, in his opinion it had not been cold enough to produce frazil ice — that it had never been observed below the St. Mary's River, 300 miles to the north. Mr. Brown said that Mr. Sloan had told him of a man who claimed that the Clinton River, a small tributary of Lake St. Clair, had "frozen solid" in about 1875, but that there were no flow observations on the Clinton River available prior to the 1930s. Mr. Rains observed that frazil ice always seemed to be a localized phenomenon and he could not visualize it extending for any distance along rivers as large as the St. Clair or Detroit.

Bristor wondered if it could be a combination of flow and some kind of ice — if it were possible to correlate flows, temperatures, and ice in past records to see if the present combination had ever existed, and what the outcome had been, i.e., had the rivers adjusted themselves? Mr. Brown said it would take 3 weeks just to correlate flows and temperatures, but that reports of ice had been made only for a few years, and even these reports were quite general in nature.

The secretary returned with the news that the Detroit Police reported water over the curb at the corner of Eight Mile Road and Woodward Avenue. Mr. Ash called the Detroit Water Department, who said a broken water main had caused the flooding at the corner, but that the main burst because the Army Engineers had raised the lake level of the inlet on Lake St. Clair so high that the main could not withstand the pressure.

Breathing a sigh of relief, Jack said that maybe the bottom of the rivers should be considered, too, and he asked if a shoal of bottom material could have formed. Mr. Ash said that while there was some loose material in the St. Clair River, the bottom was relatively stable, and that the Detroit River was pretty much of a hard-bottom river. Both were considered to be clear-water streams, not at all like meandering alluvial rivers.

At this point the aerial observer called Col. Bristor. The St. Clair and Detroit rivers appeared to consist of solid ice upon which the irregular chunks from the previous freeze were piled. Lake St. Clair was somewhat the same around the edges, but was more of a solid sheet of ice toward the center. While the ice was irregular and spotty, there was no evidence of

any obstruction, ice bridge, ice dam, or any other localized pile-up of ice.

No one could offer any reason for the rising of the river system. Bristor then asked if changing the flow in the St. Mary's River would accomplish anything. Mr. Brown replied that the storage in Lake Michigan-Huron was so vast that even complete stoppage of the St. Mary's River would have no measurable effect for several weeks. Similarly, reducing the flow of the Niagara River at the outlet of Lake Erie would not measurably raise the level of Lake Erie for over a month or two.

Bristor asked Mr. Darcy, who had mentioned icebreakers, what the icebreaker situation was. Mr. Darcy replied that the Coast Guard had icebreakers at Detroit and Toledo. Bristor then called Commander Slossor of the Coast Guard district in Cleveland who furnished the following information.

While the Coast Guard icebreakers were *not* under the operational control of the Army, the District Commander would be willing to dispatch the icebreakers for flood-control purposes upon request. The icebreakers were maintained in a high state of readiness, and could cast off in a matter of minutes. Experience during similar ice conditions indicated that the icebreakers would have no trouble maintaining passageway, that they would not encounter any obstructions, and that probably what would happen would be that, with no water traffic behind the icebreaker, the ice would flow together and eventually consolidate by freezing. The icebreakers were expensive to operate, and there were other potential users, so that a request to utilize them should not be undertaken lightly. He had no opinion as to what good the icebreakers would do.

The staff of the two districts expressed a definite opinion about the icebreakers, which Jack summarized as follows. The icebreakers could not alter the flow in the rivers. The elevations of Lake Huron and Lake Erie controlled that. If either surface ice or frazil ice were obstructing the flow of the rivers, to the extent that they could hold back 200,000 cfs of flow, the icebreakers would have no chance of penetrating the obstruction. Use of the icebreakers would be a waste of government money, as there was no known reason as to how they could do any possible good. The icebreakers would not open a waterway through the ice through which the water could flow because the ice would close in on the path they opened. In other words, the ice would flow together shortly behind the stern of the ship and freeze once more. Any effect the icebreaker might have would be strictly local. For instance, many parts of the rivers were over a mile wide; the icebreaker was only 20-ft wide, and the propeller only 4 ft in diameter. The ice cover would dampen any waves created by the bow of the icebreaker. Turbulence from the ship pushing aside the water and ice would not be felt beyond 100 ft from the ship. This was an insignificant distance in a river more than 5,000-ft wide. The same was true of any effect of the ship's propeller, which would also be local and would affect only a small part of the cross-section of the river.

Bristor asked Mr. Brown to call Mr. Sloan to see if he could furnish any useful information or opinions. Then he asked what would happen if the water continued to rise. As there had never been any flood threat of this nature, no study of such a problem had ever been undertaken and no stage-damage curves had been prepared. From a knowledge of the area, there was a general consensus that continued rising water for another day would cause great damage.

Mr. Brown returned and said that Mr. Sloan could shed no light on the situation, but concurred in the opinions previously expressed by the group.

At this time, it was reported that the Dry Dock gage was out of action. Bristor knew he had to act.

Discussion Questions

1 Was the idea to call a meeting a good one? What else could or should Bristor have done?
2 What physical principles can you apply to the situation in this case? Do they help you decide anything?
3 What alternatives does Bristor now face, what should he do next, and why?

COLONEL JACK BRISTOR Part 3

Open-channel Flow

Upon leaving the conference room, Jack Bristor went to the telephone and asked the Coast Guard to dispatch the icebreaker docked at Detroit to proceed across Lake St. Clair and up the St. Clair River to Port Huron, and to dispatch the icebreaker at Toledo to proceed across Lake Erie and up the Detroit River to Detroit. He also requested that the crews note any obstacles encountered.

Late in the afternoon, Mr. Darcy told Mr. Bristor that the icebreaker in the St. Clair River had passed the Algonac gage and that the water started to drop as soon as the icebreaker passed. Half an hour later, a similar report was received from the lower gage on the Detroit River. By the next morning, the St. Clair River gage readings had all returned to normal. The Gibraltar gage also dropped as the icebreaker passed, but went back up later in the night. The icebreaker made another run through the lower Detroit River and the gage dropped. On the next two nights the same pattern was repeated.

The Coast Guard reported that each trip was completely uneventful, that each icebreaker had steadily plowed its way through the ice for the entire trip, that no bump or thud or jolt was encountered at any time, and that the ice flowed together and froze behind the icebreaker. Darcy went along on one of the trips, returning with no report of anything remotely resembling an obstruction in the river.

Warmer weather came and the ice disappeared. Although Jack did not think he merely "lucked out," he still does not understand why the water level dropped after the icebreakers passed.

Discussion Questions

1 What steps in Bristor's decision-making procedure could have been eliminated? Should they have been?
2 What further action, if any, should be taken with regard to this problem? Describe pros and cons of what might be done next and how they should be resolved.

WARREN DEUTSCH

Part 1

Design of a Satellite-control Instrument Panel

He isn't really a human-factors engineer, but the control panel he was asked to recommend photosensors for looks very awkward to Mr. Deutsch. What could make it better?

In January, 1968, Warren Deutsch, head of the Value Engineering Department at Philco-Ford in Palo Alto, California, was helping a coworker, John Chavonec, find where to buy photosensors* for an instrument panel in a satellite-control ground station. "John and I discussed the situation," said Mr. Deutsch, "but I couldn't formulate a clear mental picture of exactly what it was he needed. I felt I'd be more help if I could see the instrument panel, and also my curiosity was getting the best of me."

Mr. Chavonec led Mr. Deutsch to a room that doubled as a customer showroom and a testing area for the Human Factors Department. There he pointed to a full-scale mockup of the instrument panel, explained its operation, and told why he felt photosensors were needed.

"Once John had explained the problem," said Mr. Deutsch, "I knew I'd have no trouble locating the photosensors he needed. I'd worked with photosensors on other projects and knew several vendors who could give us the lowdown on the proper type of photosensor for this job. But I'd also become intrigued with the manner of operating the instrument panel. It struck me as being a real hassle for the operator; too many manipulations were involved meaning wasted time and effort. I felt I could design a better one."

---

* Photosensors are activated by the presence or absence of light. Equipment incorporating photosensors can take advantage of this property and can be designed so that inputs can be "read" by photosensors and this information can be relayed.

Description of Mr. Deutsch's Job

Mr. Deutsch [see his résumé, Exhibit 1 — Warren Deutsch (Part 1)] described his position within Philco-Ford as relatively autonomous, inasmuch as value engineering was an activity that cut across other departments. He spent much of his time working on engineering problems he had discovered around the plant. The Value Engineering Department didn't evaluate all the equipment that was put out by Philco; instead, when Mr. Deutsch or another member of the department observed something that looked like it might be improved they would endeavor to determine if a better method did indeed exist. "To be successful in my line of work," said Mr. Deutsch, "it is essential that you have a good general knowledge of techniques stemming from a wide background, a natural inquisitiveness, and a good imagination." Excerpts from the Value Engineering checklist of another company appear in Exhibit 2 — Warren Deutsch (Part 1). An excerpt from government procurement regulations describing value engineering appear in Exhibit 3 — Warren Deutsch (Part 1).

In response to the question of whether or not he followed procedures as outlined in value engineering books,* Mr. Deutsch said, "I don't work from checklists or a regimented procedure, but rather start at one point and proceed by a process of association. For example, there's the well-known problem of machine failure due to the breakdown of components because of excessive heat. In testing equipment, then, it would be nice if there were an easy way to detect this excessive heat in particular components. I was reading an article in *Life* the other day about paints that diffuse into materials and change color with dif-

---

* For a description of these procedures, see Lawrence D. Miles, "Techniques of Value Analysis and Engineering" (New York: McGraw-Hill, 1961), p. 267.

ferent degrees of temperature. I associated this characteristic of the paint with the component breakdown problem. Now I'm investigating the possibility of using the paint to detect which machine parts will fail because of heat. In working through association like this I can end up anywhere, but I've found that there's no such thing as a blind alley. Any end result that is rejected yields clues that are indications of doing things another way. These clues provide a starting point.

"Any problem invariably gets bogged down some time or other; I may reach a point where I can't think of solutions or I may simply get tired of working on it. There's also a lot of delay involved in clearing up red tape in the mechanics of obtaining needed materials and in testing the efficiency and customer appeal of a product. For example, as a hobby at home I'm designing a better potter's wheel. I've sent a number of prototypes to schools in the area where they're being tested. I'll use the feedback from these tests to work out any bugs that develop, but until this information comes back, which might take weeks, I'm stymied. So I've found that I produce best when I have three or four problems going at the same time — when one gets waylaid I can take up another.

"A book will tell you that a value engineer is supposed to come up with (1) a better product at the same cost; (2) an equally usable product at a lower cost; or (3) a more efficient product at perhaps a slightly higher cost. I suppose if you took a distant view of my work you would find that I satisfy these requirements. But I'm rarely conscious of them. I consider my job to be problem solving."

Satellite Control

Man-made satellites are intended to orbit the earth and send back information useful to scientists. A limiting factor is their lack of power for sending information continuously. Consequently, provisions are made for sending it intermittently. Ground-control stations are thus equipped with instrumentation for sending commands to a satellite, which control its functions. For example, to learn the temperature of the satellite, the operator at the control station can push a particular button, pull a switch, or turn a dial to command the satellite to send back a reading. With a similar manipulation the operator can open the satellite's solar vanes. For the typical satellite there are about 100 such functions that might be controlled at the ground station by an instrument such as the one with which Mr. Deutsch had become involved.

Philco-Ford manufactured satellite-control equipment. The Human Factors Department at the plant in Palo Alto was testing the instrument panel Mr. Chavonec had shown to Mr. Deutsch. One wall of the testing room was one-way glass, on the other side of which was the testing equipment. The Human Factors Team at Philco would simulate actual satellite control situations on the panel and observe from behind the glass how operators reacted. Thus they tried to determine the ease and effectiveness with which they manipulated the controls on the instrument panels. An illustration of a typical control room appears in Exhibit 4 — Warren Deutsch (Part 1), which shows a tracking console similar to that on which the panel of interest to Mr. Deutsch was to be mounted.

Operation of the Existing Panel

The part of the instrument panel being tested was called the Page Overlay Keyboard (POK). It was the man-machine interface in a system for satellite control [like the tracking console in Exhibit 4 — Warren Deutsch (Part 1)], and was part of a large console. It had been designed at the Philco-Ford plant in Houston, Texas.

From the operator's point of view the POK panel [Exhibit 5 — Warren Deutsch (Part 1)] consisted primarily of 32 buttons. Pushing any one of the buttons would make the satellite perform a particular function. Since the panel was to be used for a wide variety of satellites, whose functions differed greatly, provisions had to be made so that the panel could be adapted to each satellite. The Philco engineers in Houston had dealt with this problem by devising an overlay system. "Books" [Exhibit 6 — Warren Deutsch (Part 1)] containing 10 "pages" with holes corresponding directly to positions of buttons on the panel had been constructed. Each satellite would have a particular book associated with it. For a given satellite a book was to be selected by the operator and inserted in the slot in the instrument panel. Photosensing devices in the slot would tell the machine which book was in place (therefore, which satellite was in operation). A particular page of the book would be selected, placed over the buttons, and held down by a bar [Exhibit 7 — Warren Deutsch (Part 1)], which also contained photosensing devices that told the machine what page was in place. The function associated with each button was determined for the machine by which book and page were in place.

Deutsch's Reaction to the POK

"When John explained the operation of the POK," said Mr. Deutsch, "I immediately thought I saw opportunities for improvement in both its mechanical and human factors. I stood beside the console and flipped through the motions needed to operate a satellite. In particular, I had uneasy feelings about the pages being contained in a book, and about the fact that the photosensor bar had to be lifted and replaced every time a page was turned. I checked the customer specification and found that the POK as designed fulfilled the stated requirements.

"One of the first things that occurred to me was that the book system provided for 320 functions (32 per page, 10 pages per book). My experience with satellite programs made me think that this number was too high. I made a few calls around the plant and got some copies of satellite programs that Philco had recently been involved with. From these it appeared that no one satellite would ever be required to perform more than 100 functions, and that all satellites had about 60 functions in common. These values, though far from accurate, seemed to me in the right ballpark, and I felt they substantiated my belief that the book system provided for more future growth than necessary. That afternoon I came up with the idea of having the pages contained in a well-like structure [Exhibit 8 — Warren Deutsch (Part 1)]. The operator had merely to select a page from the well and place it over the buttons. That way the book system was eliminated. However, there still seemed to me to be quite a lot of wasted time and motion, and I was sure there were other possibilities to explore. I decided I would find out all I could about the POK.

"The Human Factors Department had been working with the POK, and a good friend of mine, Stu Langdoc, was on that team. I had a hunch he might be able to give me some information. As it turned out he had become interested in the POK as a result of attending a briefing at ASCO earlier in the year, at which it was discussed. He learned that once the satellite was in operation the pages on the keyboard could not be turned. So it seemed to him that the rest of the pages in the book were superfluous. This led him to the idea of having only a single page overlay. Stu explained that in satellite control the operator uses the panel relatively infrequently and therefore he felt that the major design criterion to be met

was simplicity of operation, and what could be simpler than a single page? Stu had carried out a preliminary study to determine exactly how many functions were required of a satellite. He found that there were 11 functions common to all satellite programs and about 40 that were variable. He thus felt, and I concurred, that the size of a single-page overlay could be one that allowed for 51 functions. But in deference to future growth a switch matrix of 17×6 was agreed upon. Stu's information and ideas intrigued me and I decided to investigate a single-page system. There was no doubt in my mind that with a reduction of 218 computer-controlled functions there would be accompanying cost reductions in programming and manufacturing, especially if a single-page or overlay system that could be easily changed could be devised. The first question was where to start."

Discussion Questions

1 Describe possible definitions for Mr. Deutsch's problem and possible alternative solutions.
2 What is the company getting when it hires Mr. Deutsch? Explain what he seems especially well-equipped to contribute and contrast this with what the company might need from others.
3 Tell how you would proceed at the end of the case and carry out your approach as far as you can.
4 Comment on the statement "value engineering is just common sense."
5 Consider Mr. Deutsch's résumé, which appears in Exhibit 1 — Warren Deutsch.
    a How do the things he has done compare and contrast with activities other engineers in the cases are engaged in?
    b What is the purpose of this résumé?
    c What does the résumé not tell that might be relevant to its purpose?
6 Evaluate Mr. Deutsch's background as an applicant for a job at Task Corporation, or how Dick Rigney would react to him (Chapter 4).

Exhibit 1 — Warren Deutsch (Part 1)

RÉSUMÉ

Warren A. Deutsch
823 Rorke Way
Palo Alto, Calif. 94303

Married, 4 children   Age: 40   5′8″   180 lbs.

Education: Bronx High School of Science, 1947
             University of Chicago, Physics, 1951–1953

Work Experience:

1963–Present

Philco-Ford Corporation, Western Development Laboratories, Palo Alto, California

Value Engineering and Advanced Techniques Manager, Aerospace Ground Operations

As Value Engineering Manager, responsibilities include the initiation and evaluation of Value Engineering ideas, selecting engineers/specialists for fact finding teams, control of funding for investigations, consulting services, and materials appropriate to the area of interest, preparation and presentation of final Value Engineering Proposals to management and the customer (USAF, NASA, etc.). Directly responsible for development of a floundering program into a profit center which last year earned fees equivalent to 1.5 million dollars in new business.

Recipient of an award presented by the Air Force in recognition of outstanding contributions to their Cost Reduction Program. The Advanced Techniques office acts as the focal point for the collection and dissemination of various types of information that for the most part represents ideas and techniques developed by both private and government-sponsored programs. Categories of information collected include medical, mechanical, electrical, electronic, chemical and computer programs. The proper dissemination of this information requires a wide range of knowledge and interest as well as a constant awareness of divisional, program and individual needs and interest. In addition, it requires the ability to adapt information derived of one discipline to the requirements of another. Engineering Specialist, Product Assurance, Reliability Department Staff to the Manager. Responsible for special programs and projects including use of computer collection and processing of reliability data in conjunction with a program designed to integrate purchasing, vendor, quality, material, where used and use data; use of infrared microscope as a quality control tool, as a reliability test monitoring instrument, and operational fault-finding monitor. Provided technical assistance to other departments and participated in inter-departmental study teams as a representative of the Reliability Department.

1963

Israel Aircraft Industries, Lod, Israel

Consultant to the Associate Director

Developed an organizational plan that reflected the functional relationships of the electronic division's various activities and anticipated a rate of growth approximating 100 per cent per year for several years. Advisor to the departments of Engineering, Manufacturing, Quality Control and Material on questions of function and procedures.

Demonstrated that unused materials after some modification could take the place of scarce, expensive or unobtainable materials. For example, fiberglas insulation and burlap became modular ceiling tile for acoustical treatment in noisy areas. Aircraft scrap was fashioned into playground equipment as the company's contribution to the community.

1959–1963

Philco Corporation, Western Development Laboratories, Palo Alto, Calif.

Manager, Failure Analysis Laboratory

Proposed the establishment of a Physics of Failure Laboratory as a natural adjunct to Reliability's function and necessary for a better understanding and control of the physical factors affecting Reliability. Developed complete plan for its budget, layout, instrumentation and personnel. Assigned the responsibility for carrying out the plan; construction, installation of instrumentation, and selection of personnel. In addition became a registered industrial radiologist and health physicist. Performed tests necessary to receive State approval of facility and military approval of procedures. Appointed Manager responsible for (1) analysis of all part failures which occur in the development, fabrication, test shipment and operation of Philco WDL satellite-systems equipment, and (2) developing a test and analysis program for determining the mathematical relationships between environments and failure-causing changes in part materials — the ultimate goal is reliability prediction based on the calculation of part behavior in a given environment rather than the projection of empirical data. Specializing in the analysis of semiconductor failures and the development of advanced failure-analysis techniques. Designed and built a transistor case cutter that in no way introduced contaminates into or injured the devices. Systems then used by industry injured the device or were time consuming and in some instances used acids. The cutter could be used by a technician and usually did not take longer than three minutes to complete the operation.

Was instrumental in discovering a fault in the manufacturing techniques used in producing a monolithic glass semiconductor diode. Defect, discovered by the use of x-ray, would have caused a number of satellite equipment failures.

Examination of hydraulic fluid from failing servo-valves disclosed the source of contamination as coming from another system component. This work led directly to a redesigned part and a procedure that provided a check on the condition of the entire system.

Engineering and Administrative Assistant to Manager Reliability Assurance Department

Liaison between Reliability Assurance Department and engineering laboratories to assure effective integration of reliability and engineering efforts. Responsible for the direction of engineering efforts within the Reliability Assurance Department; recommendation and implementation of procedural and organizational changes to increase the effectiveness of the reliability program.

Senior Engineer, Reliability Assurance Department

Initiated study of high-altitude environments similar to previous investigation for Admiral Corporation.

1959

Lockheed Aircraft Company, Missiles and Space Division, Sunnydale, California

Senior Instrumentation Engineer

Reviewed procurement requirements for major items of environmental equipment; prepared procurement specifications. Performed technical evaluation of manufacturers' bids and liaison between manufacturers and engineering departments. In many cases made major modifications to engineering designs and introduced a method by which purchasing could evaluate bidders.

1959

Armour Research Foundation, Chicago, Illinois

Research Associate

Evaluated nondestructive vibration testing of miniature and subminiature electronic tubes as a method for predicting tube reliability. Determined that the operational behavior of commercial test equipment varied sufficiently to prevent correlation of results obtained from different test facilities. Test program amended to establish Armour as the sole test facility.

1958–1959

ECM Laboratory, Admiral Corporation, Chicago, Illinois

Senior Engineer, Physicist

Surveyed existing reliability test and evaluation facilities; recommended improved use of existing facilities and new facilities necessary for establishing a more effective reliability-evaluation program.

Reliability Evaluation Laboratory Director

Appointed Laboratory Director responsible for establishing new Evaluation Laboratory in accordance with recommendations resulting from facilities study.

Studied high-altitude environments for the purpose of defining environmental conditions more precisely and reducing, if possible, the 'environmental support' equipment required for high-performance aircraft systems.

Results of this study were used: (1) as a basis for writing specifications for equipment in the research and development stage, and (2) to establish realistic environmental test requirements for reliability evaluation.

1954–1958

Hallicrafters Corporation, Chicago, Illinois

Director, Reliability Laboratory

Responsible for planning and supervision of reliability testing; program included test of Atlas Missile System fuel and hydraulic-system components and a variety of ECM equipment in the IF, IR, and visible spectrums.

Engineering and Administrative Assistant to Reliability Laboratory Director

Responsible for the analysis and evaluation of reliability test data and the preparation of evaluation reports. Prepared costing and technical information for proposals issued in response to request for reliability testing received from other firms.

Engineer

Designed miniaturized circuits for Electronic Counter Measures equipment. Circuits developed included a sweep generator, IF amplifier, and audio amplifier which were considerably smaller than contemporary circuits of comparable electrical characteristics and reliability.

1953–1954

Institute for the Study of Metals, University of Chicago

Research Technician

Studied the crystalline structure of various Boron steels. Participated in an extended study of Alpha and Beta brasses to determine the cause and structural significance of a metal-temperature 'plateau' which occurs when these materials are exposed to a gradual temperature increase. Performed micro-hardness tests and x-ray analysis of test specimens and statistically reduced data obtained. Designed a photographic specimen-jig for positioning specimens during x-ray analysis.

Exhibit 2 — Warren Deutsch (Part 1)   Partial value engineering checklist (developed by another company for use by its Design Review Committee)

1. Have the customer's specifications been subjected to a systematic review to determine whether they require more than is needed? Do they include excessive cost-producing requirements relative to high temperature, shock, vibration, or other environments?

2. Have engineers challenged the design of individual components with respect to economy and suitability of manufacture? For example, shall a part be machined, cast, forged, or welded?

3. Have quantities of parts required for the development program and their costs been considered?

4. Have designs been challenged with the objective of recommending better designs when they represent lower cost while retaining the required functions?

   a. What is the function of a given part? Is it required?
   b. How much does the present design cost?
   c. What else will perform the same function? At what cost?

5. Have standard off-the-shelf items been used in the design wherever reliable items are available? Have potential vendors been consulted for alternatives or modifications that would reduce costs?

6. Does the design represent optimum simplicity commensurate with functional requirements? Can any part be eliminated or combined with another part to reduce the total number of parts and cost?

7. Has relative workability and machinability of materials been considered? Have specifications been reviewed to eliminate unnecessary requirements? Can the design be modified to use the same tooling for right and left handed or similar parts? Are all hand operations essential? Could furnace brazing be substituted for manual welding, for instance? Do hole sizes use standard drills? Has deep hole drilling been minimized?

8. Have experienced engineering and manufacturing specialists been consulted where they can likely help with the design? Have tool and manufacturing engineers reviewed the design before release of prints?

9. Are all machined surfaces necessary? Could less expensive finishing be used? Has the required finish been specified at the proper stage of completion to minimize damage during subsequent handling?

10. Are all parts designed for assembly at the earliest possible time considering that assembly costs increase as system build-up progresses? Have state-of-the-art bonding techniques been considered for applications requiring riveting, bolting, and their associated finishing operations?

Exhibit 3 — Warren Deutsch (Part 1)

Part 17—Value Engineering

1-1701 Policy.

(a) *General.* Value engineering is concerned with elimination or modification of anything that contributes to the cost of an item but is not necessary to required performance, quality, maintainability reliability, standardization, or interchangeability.   Value engineering usually involves an organized effort directed at analyzing the function of an item with the purpose of achieving the required function at the lowest overall cost. As used in this Part, "value engineering" means a cost reduction effort not required by any other provision of the contract.   It is the policy of the Department of Defense to incorporate provisions which encourage or require value engineering in all contracts of sufficient size and duration to offer reasonable likelihood for cost reduction. Normally, however, this likelihood will not be present in contracts for construction, research, or exploratory development.   Value engineering contract provisions are of two kinds:

    (i) value engineering incentives which provide for the contractor to share in cost reductions that ensue from change proposals he submits; and

    (ii) value engineering program requirements which obligate the contractor to maintain value engineering efforts in accordance with an agreed program, and provide for limited contractor sharing in cost reductions ensuing from change proposals he submits.

(b) *Processing Value Engineering Change Proposals.*   In order to realize the cost reduction potential of value engineering, it is imperative that value engineering change proposals be processed as expeditiously as possible.

1-1702 Value Engineering Incentives.

1-1702.1 *Description.* Many types of contracts, when properly used, provide the contractor with an incentive to control and reduce costs while performing in accordance with specifications and other contract requirements. However, the practice of reducing the contract price (or fee, in the case of cost-reimbursement type contract) under the "Changes" clause tends to discourage contractors from submitting cost reduction proposals requiring a change to the specifications or other contract requirements even though such proposals could be beneficial to the Government. Therefore, the objective of a value engineering incentive provision is to encourage the contractor to develop and submit to the Government cost reduction proposals which involve changes in the contract specifications, purchase description, or statement of work. Such changes may include the elimination or modification of any requirements found to be in excess of actual needs regarding for, example, design, components, materials, material processes, tolerances, packaging requirements, or testing procedures and requirements. If the Government accepts a cost reduction proposal through issuance of a change order, the value engineering incentive provision provides for the Government and the contractor to share the resulting cost reduction in the proportion stipulated in the value engineering incentive provision.

Exhibit 4 — Warren Deutsch (Part 1)

Exhibit 5 — Warren Deutsch (Part 1)

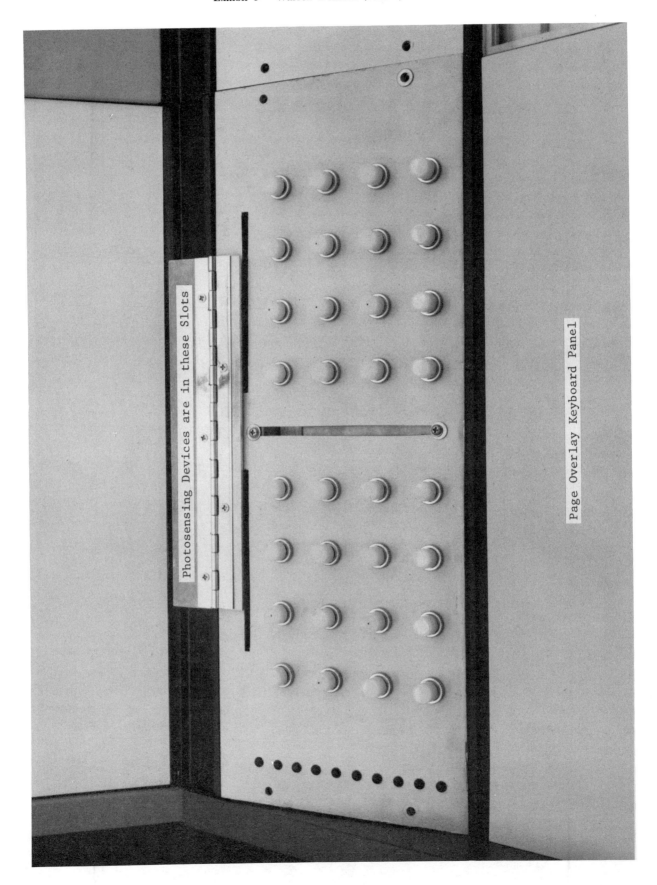

Exhibit 6 — Warren Deutsch (Part 1)

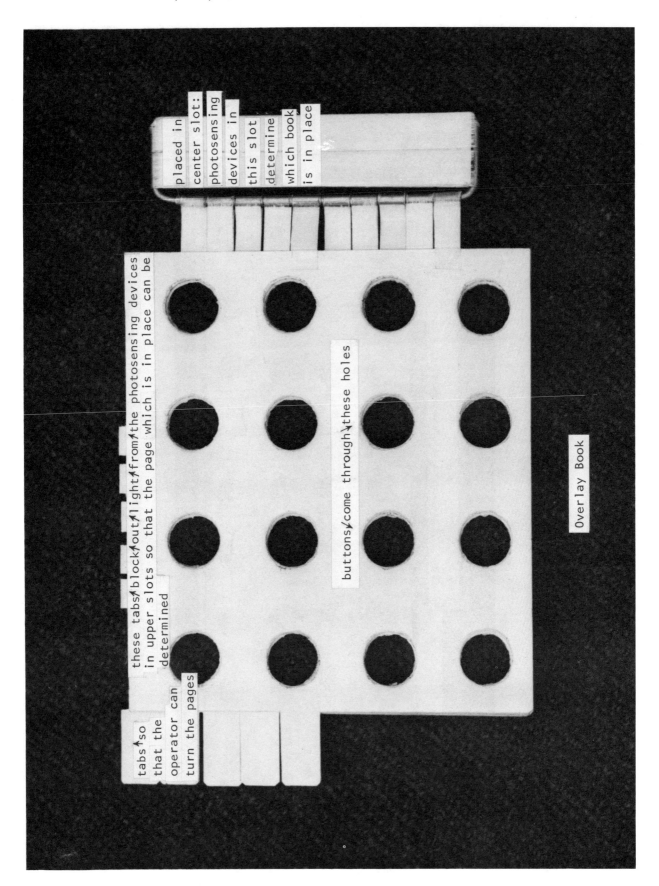

Exhibit 7 — Warren Deutsch (Part 1)

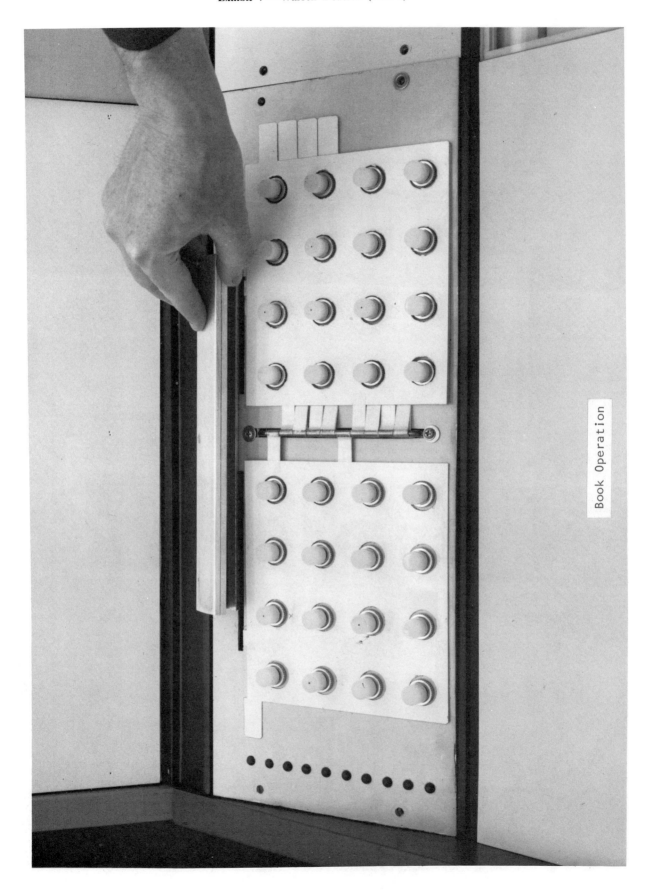

Book Operation

Exhibit 8 — Warren Deutsch (Part 1)

A number of pages are held in the
"well" and are indexed so that the
operator can choose the appropriate
one easily

grooves

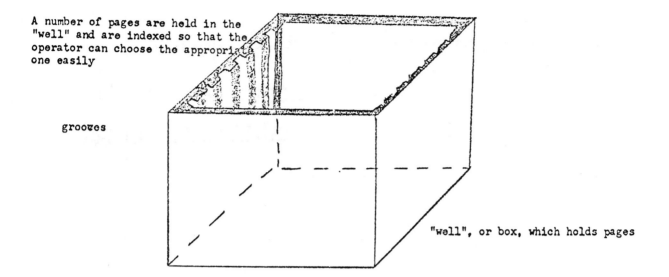

"well", or box, which holds pages

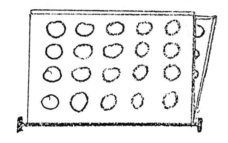

rod slides down grooves

Mr. Deutsch's Well Idea

WARREN DEUTSCH

Part 2

Design of a Satellite-control Instrument Panel

Mr. Deutsch had settled on a single-page overlay system for the controlling panel. He found it helpful in following through the idea to discuss it informally with Stu Langdoc and other members of the Human Factors Department who understood the specifications to be met. "For instance," said Mr. Deutsch, "Stu pointed out that the best method of operation was pushing buttons, because the operator would feel more at home with the system. Because of this suggestion, because the POK system had used buttons, and because I was more familiar with buttons, I decided to work with push-button-type switches on the keyboard."

A number of problems now arose. Nomenclature had to be associated with each button so that the operator could tell what function would be performed by the satellite when he pushed a particular button. In the POK system the nomenclature had been written below each hole in a page so that when the page was in place the nomenclature would appear below each button. But the POK had used only 32 buttons, and Mr. Deutsch had deduced that a single-page overlay system might require as many as 110 buttons.

"I wasn't sure if there was a specific area on the console to which the keyboard would be confined," said Mr. Deutsch, "but I knew that if the keyboard became too large it would make it tougher on the operator. So I ruled out the possibility of putting it below the buttons, because this would make the keyboard too big. The answer was fairly obvious to me; somehow put the nomenclature on the overlay covering the buttons. I wasn't sure how I could do this but I stored the idea in my mind with intentions of working out the 'hows' later."

Mr. Deutsch worked through what he described as a process of association to come up with his next idea. In the POK system with the nomenclature below the buttons there could be holes in the overlay and the buttons could be pushed directly. "But if

the new single-page keyboard was to be made up of buttons to be pushed," said Mr. Deutsch, "and if the overlay was to have nomenclature covering the buttons, then the overlay had to be flexible so you could push 'through' it to activate the button. I decided to try using an elastic membrane to cover the buttons. I talked to a number of vendors to see what they had and decided vinyl would be the most functional, because the other types were too expensive or couldn't take printing."

The keyboard that Mr. Deutsch had decided on was to have roughly 110 buttons. However, for a given satellite program the number of functional buttons was likely to be much smaller. For example, if 50 operations were required of a satellite, there would be 50 functional buttons and 60 "dead" buttons left over. The 50 functional buttons would normally be separated into functional areas. For example, 20 buttons might govern maneuvers of the satellite while the other 30 might pertain to its sending information to the control station. These two groups would represent the two major functional areas. These areas in turn could be separated into minor, smaller, more specific functional areas; i.e., of the 20 buttons governing maneuvers, 5 might control mechanisms governing speed, 5 might apply to rotation, and 10 might apply to the orientation of instruments.

With this information in mind, Mr. Deutsch got together with Mr. Langdoc to discuss the design of the vinyl overlay. "We just started talking," said Mr. Deutsch, "remembering that our goal was to make the keyboard simple to operate; and the final design 'fell out.' We decided that the best way to keep it simple would be to make a cutout to accompany each vinyl overlay. This cutout would be placed over the overlay, and only those buttons in the cutout region would be exposed. The cutout regions would correspond to the functional areas [Exhibit 1 — Warren Deutsch (Part 2)]. Another thing we decided was that we wanted color-coding of the functional areas. That is, if 5 buttons operated mechanisms controlling motion of the solar vanes, we wanted these buttons to be all one color — say blue. Each minor functional area would be colored differently [keyboard appears in Exhibit 2 — Warren Deutsch (Part 2)]. In deciding how to do this we immediately ruled out coloring the overlays when they were processed, because we thought it would be too expensive. Instead we knew we had to adapt the overlays once we had gotten them from the manufacturer. I thought about it and came up with an idea that would also allow for putting the nomenclature above the buttons as I wanted. The overlay could be designed so that plastic inserts could be placed below each area on the overlay corresponding to a button [Exhibit 3 — Warren Deutsch (Part 2)]. The nomenclature would be printed in black on the first insert, which otherwise would be clear and transparent. A translucent solid colored insert would go below this. The vinyl overlay would be clear and transparent. The operator would see through the overlay and the black nomenclature would appear superimposed on a solid color.

"I was happy with this idea, but Stu pointed out that at any one time there would be functional areas that would be temporarily inactive. He felt it would be nice if their colors could be temporarily obscured and appear as white to the operator. For example, a typical satellite program might have a sensing device — call it the XYZ — contained within the satellite that, when exposed to the atmosphere, obtained several bits of information and sent them back to the control station. Two different functional areas are represented here. A red [arbitrarily selected] area would control mechanisms of the satellite and thus contain a button that would operate a mechanism that would expose the XYZ to the atmosphere, as well as another to bring it back into the satellite again. The other functional area, a green one, would control sending the information from the XYZ and would contain buttons for getting the temperature, pressure, and the percent oxygen in the air from the XYZ. When the XYZ is inside the satellite, it is incapable of obtaining information, and thus the green functional area would be inactive. Because of this we would want the green area to appear temporarily as white. But when the button in the red area is pushed, which exposes the XYZ to the atmosphere, the green area becomes active and the operator should see it as green rather than white.

"I thought this over for a while and finally came up with a scheme that would meet Stu's demand. The vinyl overlay could be translucent white rather than clear. Each button would have a light in it controlled by the computer. When the light was on the color and nomenclature would show through the overlay; when the light was off the button would look white. Suppose the XYZ is in the satellite. The

lights in the red area would be on while those in the green area would be off. The red functional area would be lit red, but the green functional area would be temporarily white. Suppose then that information is needed from the XYZ. The operator looks into the red area for the appropriate button, pushes the one exposing the XYZ to the atmosphere. When he does this the computer turns on the lights in the green area, the green color shows up in place of the white, and the operator can now read the nomenclature of the buttons in that area."

Just as with the POK system the computer had to be able to "read" which overlay was in use. "There was an obvious way to do this," said Mr. Deutsch, "stemming from the POK system. Each overlay would be coded, and sensing devices on the panel would read this coding. I decided that this coding could go on the edges of the overlay, while the sensing devices could be positioned appropriately on the panel [see Exhibit 4 — Warren Deutsch (Part 2)]. This idea seemed very reasonable and I decided to stay with it. I immediately rejected the use of photocells for the sensing devices. They're double-active devices, meaning that the light in the cell could burn out or the light receiver could go on the blink. In either case the system would fail to read the overlay properly. Another thing that led me to reject photocells was that amplifiers were needed to raise the current from the photocell. Several other sensing devices came to mind, including the microswitch and the magnetic reed switch.* I'd seen these types of switches in other projects, and was fairly familiar with them. I didn't know if either would be practical for our design, so I took the magnetic reed switch, built a small mockup, and tested it. It worked fine. The reed switches could be imbedded in the console with a magnet next to each, biasing it to the inactive position. Each overlay would have pieces of iron foil imbedded strategically in its edges. When the overlay was put in place, the switches that were under the iron foil would have their magnetic field shunted and the reeds would switch to their closed positions. Thus by properly designing the placement of the switches and foil and then programming this information into the computer, the computer could "read" which overlay was in place. As an added benefit, I found that the magnet had sufficient strength to hold the overlay in place."

Another specification that occurred to Mr. Deutsch was providing for the prevention of pushing more than one button at a time. The buttons were to be

* A magnetic reed switch [Exhibit 5—Warren Deutsch (Part 2)] has a reed that can be magnetically biased to an open or closed position (that is, either short-circuited or open-circuited). When another piece of iron comes near, it shunts the magnet to the different bias (that is, if open originally it becomes closed when iron is brought near). A microswitch is one that is activated by physical application of very light pressure.

located very close to each other, so it was possible that the operator could accidentally push more than one button at once. "In the past I've often come across the use of interlock systems," said Mr. Deutsch. "They are commonly used in devices like typewriters where only one of many adjacent buttons is to be pushed at any one time. But I knew that this interlock method was fairly complex. A lot of hardware is needed and it is expensive. I decided that I'd try and think of a better method.

"I thought about the problem and after a while came up with the idea of putting a barrier around each button. If you were pushing a particular button and your finger slipped it would hit the barrier and not another button. In exploring this idea further I thought of the old-fashioned eggcrates. These were made up of wood slats arranged in honeycomb tiers, and I thought that a similarly constructed tier, when placed around the buttons on the keyboard, would prevent the pushing of more than one button at a time, just as the tiers in an eggcrate keep eggs from hitting each other and breaking. I reviewed available literature and found that several companies had developed this barricade idea and had models for sale. I followed up these leads, got literature from the companies, and reviewed the specifications of each to see which of the barriers would be most adaptable to the panel. Several seemed suitable, so I selected one and examined it carefully. It was constructed of slats [depicted in Exhibit 6 — Warren Deutsch (Part 2)], which were held together by long, hollow rivets. In order to place the matrix (i.e., the company that made it called the barrier configuration a "matrix") in a console, a frame was built onto the perimeter of the matrix and this in turn could be screwed to the console. It seemed to me that this frame might present complications, in that when the overlay was placed over the matrix, the frame would cause it to be slightly raised from the console. This would make it difficult to provide for a proper connection between the iron foil on the overlay and the reed switch in the console. I tried to think of a way to modify the matrix so that the overlay could hit flush

with the console. My first thought was to somehow get rid of the frame, as it was causing the problem. It would be easy to do without the frame, but in doing so I had to create another way to secure the matrix to the console. I thought about it for a while and got a brainstorm — the hollow rivets tipped me off. Thin metal rods could run through the rivets and bracket to the console's wall below the surface. This would secure the matrix and allow for a flush interface between the matrix and the console surface."

Mr. Deutsch presented this idea to the company that manufactured the matrix and asked if they would modify it in the above manner. The matrix company reviewed Mr. Deutsch's idea, decided to modify it for him, and also decided to offer it as a design for future customers. Mr. Deutsch pointed out that at this stage the investigation was nowhere near complete, and that there would be a continuing evolution of the design based on factors not yet fully evaluated, such as cost, ease of manufacture, reliability, and ease of operation.

The program manager on the project added that, although the new idea might be useful on future projects, it was expected that the POK would still be used on this one. He added also that planning information, which Mr. Deutsch had earlier not been aware of, supported retention of a large number of control functions (button alternatives) to allow for future growth of the system.

Discussion Questions

1 Evaluate Mr. Deutsch's design and compare it to others suggested by yourself and the rest of the class.

2 If instead of being on salary Mr. Deutsch had been working for them as a consultant on this project, how should he compute his bill at this time? How should he go about predicting what his bill to the company should be for further work on the project?

Exhibit 1 — Warren Deutsch (Part 2)

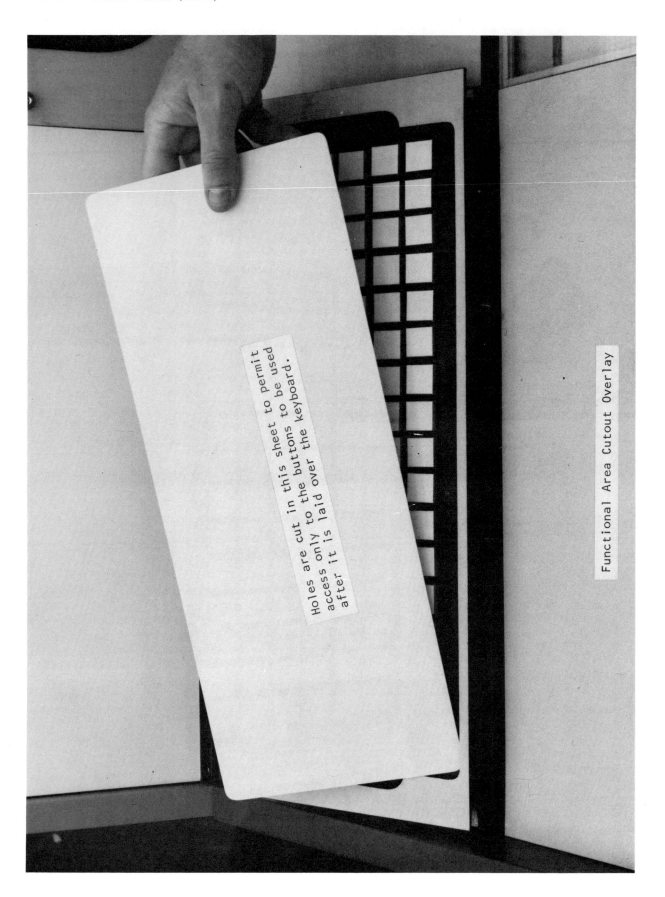

Holes are cut in this sheet to permit access only to the buttons to be used after it is laid over the keyboard.

Functional Area Cutout Overlay

Exhibit 2 — Warren Deutsch (Part 2)

Exhibit 3 — Warren Deutsch (Part 2)

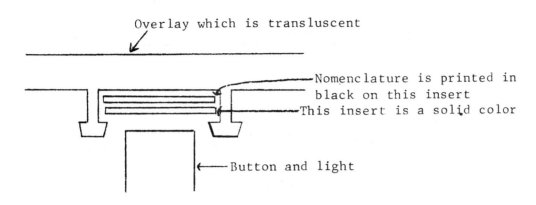

Side view of vinyl overlay and
   plastic inserts below

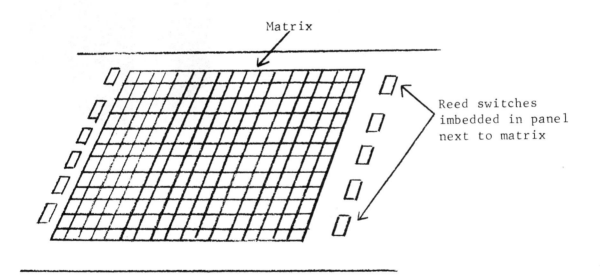

Top View

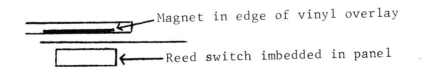

Side View

Exhibit 5 — Warren Deutsch (Part 2)

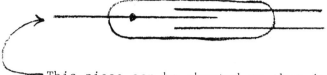

This piece can be shunted so that it contacts one or the other of the components so that it is either open or closed.

Magnetic Reed Switch

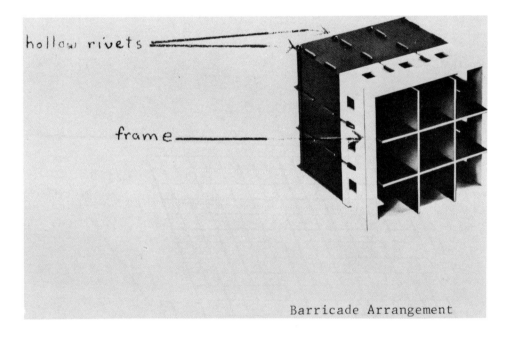

Barricade Arrangement

KEITH McFARLAND

Part 1

Removing Pincushion Distortion in a Cathode-ray Tube

Keith faces an esoteric electronics problem. Students without advanced training won't be able to cope with it, analytically, but they can try to figure out what general form the engineering approach to a problem of this sort might take.

"We spent the first three weeks of March, 1966, trying to debug the system," said Keith, "but nothing we tried seemed to work. If we got the picture to focus, we couldn't get rid of the 'pincushion.' When we straightened out the pincushion, we lost the focus again. Time was really running out on us. Our system had to meet specifications by a certain date or we would have to pay a penalty fee for each day we were late. By that third Friday, I'd decided it was no go. We were up a blind alley. I wanted to try another approach."

Keith McFarland, an Electrical Engineer in the Link Group of General Precision Corporation, was talking about Jet Propulsion Laboratory's (JPL) video film recorder. JPL had subcontracted development of the recorder (part of the Spacecraft Television Ground Data Handling System) to Link. Keith was assigned as project engineer to head the group that designed and constructed the recorder.

The video film recorder was to take photographs of the television pictures sent back to earth by such spacecraft as Ranger, Mariner, and Surveyor. The pictures were received as high-frequency signals, which were demodulated and then projected onto a cathode-ray tube (CRT) similar to those used in television sets. One of the specifications set forth by JPL required that the pictures be displayed on a flat-faced CRT, one that was flat to within 0.002 in. This would facilitate taking accurate photographs of the television pictures, so the photographs could later be enlarged and carefully studied for new informa-

tion. There was some difficulty however, in producing the pictures on a flat-faced CRT. The physics of a CRT employing magnetic deflection are such that there is minimal distortion only if the phosphor-coated screen has a spherically curved face. To avoid distortion, this curved face must have a radius of curvature the same length as the screen's distance from the center of the deflecting field.

A picture projected on a flat-tube face has an inherent distortion known as "pincushion." Near the center of the tube face, there is not much difference in radial distance from the center of the deflecting field to the curved or flat face, so there is little distortion. But'as can be seen from Exhibit 1 — Keith McFarland (Part 1), the farther from the center of the screen, the greater the difference becomes between the flat and spherical faces. Accordingly, a square, perfectly represented on a curved tube, becomes distorted on a flat one, the greatest distortion occurring at points farthest from the center of the screen (see Fig. 1).

Keith, who had been with Link since graduating in electrical engineering from Stanford in 1959, had previously worked with flat-faced CRT's in connection with a radar application on a project called a land-mass simulator. With the radar, however, a radial-sweep picture was obtained, and the distortion

becomes

Image on Spherically Curved Tube   Image on Flat Tube

was in only one dimension, the radial one. (Imagine the second-hand sweeping around the face of a clock and leaving shadow images as it passes.) This one-dimensional distortion was eliminated by adding a correction current to the current of the deflection coils in the CRT.

The picture to be transmitted by the JPL spacecraft, however, would involve two dimensions, so a radial-sweep correction factor would not correct the pincushion effect. "The specifications call for each picture element to be within 0.3 percent of its ideal theoretical position," Keith said. "Let me give you some idea of what this means. Picture elements on a good 21-in. television set are around 7 percent of their ideal theoretical position. JPL requires a picture over 20 times more precise than what you receive at home. With a screen like this, which is flat to within two-thousandths of an inch, the pincushion distortion can go all the way up to 20 percent at the edges. [Dimensions of the CRT tube to be used appear in Exhibit 2 — Keith McFarland (Part 1)].

"I first met this pincushion problem when I was working on the land-mass simulator. It is a common distortion problem; in fact there are companies that specialize in making corrective devices for it. During the preliminary design, I thought one of these devices would probably bring the distortion within specifications. I went to a catalog and looked up pincushion correctors. A company called CELCO is sort of the Cadillac of the pincushion correctors, so I ordered one of their models that looked like it would do the job."

The CELCO pincushion corrector [see Exhibit 3 — Keith McFarland (Part 1)] is a magnetic deflection yoke that fits over the neck of the CRT. The yoke produces a static or constant magnetic field that the electron beam passes through after it has been deflected by the horizontal and vertical deflection holes. This additional field causes an extra deflection, the effect of which is enough to straighten out the sides of the picture.

As project engineer, Keith's function was to supervise the engineering end of the project. "It's my responsibility to see that the system is working by the contract deadline. If we have trouble with one of the circuits, my job is to work with the engineer until we get the problem ironed out." Keith stated he had worked on several other circuit problems of the video film recorder while waiting for the CELCO pincushion corrector to arrive and be installed.

"This project has had its rough spots, but you learn to expect that in research and development contracts," he continued. "That's why research and development contracts are on a cost-plus basis. Things don't always work out the way you plan them. When you bid on a contract like this, you estimate a cost and a completion date, and you promise a certain degree of technical excellence: 'Our system will have such

and such an accuracy, availability, and so on.' Generally it is a trade-off among the three. Sometimes it costs more than you estimated or takes a little longer to deliver, sometimes less. We almost always get the technical excellence we aim for. It's usually the most important.

"We had some typical complications on this job. The specifications were changed a few times. Sometimes this was at our request when we found that the specifications weren't realistic. For example, JPL might specify lenses that just aren't available as off-the-shelf items. They would have to be specially developed under another contract with an optical company. Other times we felt some requirements were unnecessary or redundant. When we could show that this was the case, JPL was usually willing to change the specs.* JPL initiated some spec changes, too. Occasionally they had to request modifications because of design changes in other systems. We also got held up a few times by late deliveries from our suppliers.

"There have also been several circuit bugs. This pincushion effect has been giving us the most trouble. The CELCO pincushion corrector was installed and tested, and we got an unpleasant surprise. The corrector straightened out the sides of the picture pattern within the 0.3 percent we wanted, but then we couldn't get the picture to focus properly. The device caused spot growth and increased nonlinearity of the circuit. The result was a fuzzy picture that wouldn't meet the resolution specs.

"We magnified the picture to see what was happening. It turned out that each picture element, or pizel, instead of being a perfect dot had become cigar-shaped. We've spent about 3 weeks trying to improve the focus and correct for spot growth, but now it's pretty clear that debugging is not the solution. Time is very definitely becoming a problem at this point. Whatever we do has to be done quickly."

Discussion Questions

1 How many different approaches to the problem faced by Keith McFarland can you list?
2 Briefly describe how you would go about carrying out each of the above approaches and speculate as to the length of time it would take you to produce a satisfactory solution with that approach.
3 If you were in charge of four others with the same qualifications as yourself, and the team of four was asked to produce a solution as fast as possible, what assignments would you make? How would you manage the team?

_____
* Specifications.

Exhibit 1 — Keith McFarland (Part 1)

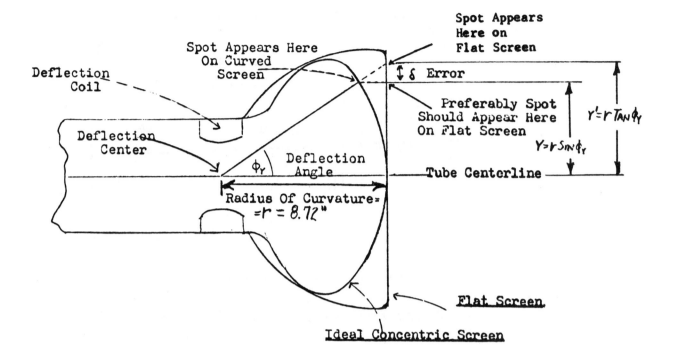

Exhibit 2 — Keith McFarland (Part 1)

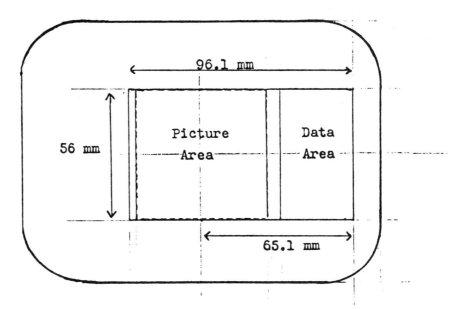

Cathode Ray Tube Face Showing Actual Display Area

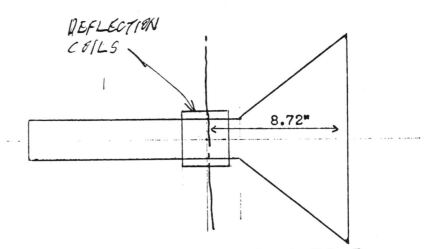

CRT Showing Distance From Center of Deflection to Tube Face

Exhibit 3 — Keith McFarland (Part 1)

**PINCUSHION
CORRECTORS
TYPES E, U, L, M**

CORRECTIVE FIELDS
FOR
CRT PINCUSHIONING

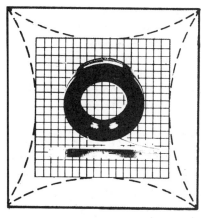

TYPE E
**Popular Precision Pincushion
Corrector**

Straight Sides to 0.25%	
Resistance	150 ohms
Current	60 maDC
25 KV with CL1119 CRT	
O.D.	4½″
I.D.	2⁹⁄₁₆″
Thickness	1⁵⁄₃₂″
Mounting Bore	$\frac{4.004}{4.002} \times \frac{1}{2}''$

CELCO Pincushion Correction Assemblies have been used by display designers since 1953 for straightening the sides of the well known pincushion pattern on the CRT face.

The static correction field produces magnetic forces that operate on the electron beam in the drift space between the beam exit of the deflection yoke and the CRT face.

The standard units are available as shown, or in combination with any CELCO Deflection Yoke or Deflectron. (See other side of this sheet.)

CELCO specializes in optimizing field correctors, deflection yokes and CRT's into complete, integrated packages for minimum spot growth, straightest sides and linearity correction. (See special notes on reverse side.)

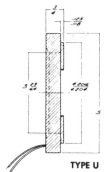

TYPE U

Ultra-Precision
Straight Sides to 0.1%
Use with CELCO C1628-3 Micro- Positioner

Resistance	100 ohms
Current	125 maDC

10 KV with 5CEP CRT

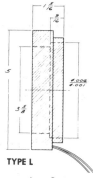

TYPE L

Low Cost
Straight Sides to 1.0%

Resistance	75 ohms
Current	300 maDC

27 KV with C5A11 CRT

TYPE M

Permanent Magnet
Straight Sides to 1%
No D.C. required
Use with Direct View Displays
on 10UP, 22CP, 5CEP or other CRT's

Standard data listed above. Other CRT's and special application on request. Call our Engineering Department.

Constantine Engineering Laboratories Company

"For the latest in the science of Electron Beam Control"

PACIFIC DIVISION
Upland, California
Area code: 714 YUkon 2-0215
TWX 714 556 9550

SOUTHERN DIVISION
Miami, Florida
Area code: 305 PLaza 1-1132

EASTERN DIVISION
Mahwah, New Jersey
Area code: 201 DAvis 7-1123
TWX 201 327 1435

Celco Catalog Sheet C-1

STANDARD PINCUSHION CORRECTOR AND DEFLECTION YOKE ASSEMBLIES

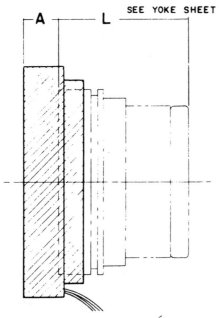

SEE YOKE SHEET

OVERALL LENGTHS		
TYPE E	A = $^{21}/_{32}$ PLUS	L
TYPE L	A = $^{3}/_{4}$ PLUS	L
TYPE M	A = $^{35}/_{64}$ PLUS	L

PINCUSHION CORRECTOR AND YOKE

TYPE E—Precision Electromagnetic Pincushion Assembly

TYPE L—Low Cost Electromagnetic Pincushion Assembly

TYPE M—Permanent Magnet Pincushion Assembly

TYPE U—Ultra Precision Electromagnetic Pincushion Assembly

These Pincushion Correctors may be used with any CELCO AY, FY, HY or HD yoke or Deflectron (Ultra High Resolution Yoke) with the housing reversed. The assembly type number becomes AYE521- for a CELCO yoke AY521- (Sheet Y3A) and a Type E pincushion corrector; FYL727- for a CELCO yoke FY727- (Sheet Y17) with a Type L corrector. HDE428- describes a Deflectron HD428- (Sheet D2A) with a Type E corrector. Your CELCO sales engineer can help.

NOTE: Spot growth and linearity are degraded with all ordinary yoke and pincushion corrector assemblies.

SPECIAL NOTE: Celco has developed special Electron-Optical equipment to minimize spot growth in conjunction with pincushion correction.

When the pincushion corrector, the deflection yoke and the cathode ray tube are considered as a unit with all components coupled, a combination may be produced which will meet almost any resolution problem for a display with straight sides.

(Consult our Engineering Department.)

SPECIAL NOTE: Although the pincushion corrector may degrade absolute linearity, straight sides are produced at all values of X-Y current through the deflection yoke. It is possible, therefore, to supply shaped waveforms to the X-Y deflection amplifiers to obtain on-axis correction and to achieve linearity correction for the combined X-Y deflection on the tube face.

These waveforms must be the reverse of those non-linearities which are produced by the CRT face geometry, the yoke and the pincushion corrector.

Constantine Engineering Laboratories Company

"For the latest in the science of Electron Beam Control"

PACIFIC DIVISION
Upland, California
Area code: 714 YUkon 2-0215
TWX 714 556 9550

SOUTHERN DIVISION
Miami, Florida
Area code: 305 PLaza 1-1132

EASTERN DIVISION
Mahwah, New Jersey
Area code: 201 DAvis 7-1123
TWX 201 327 1435

KEITH McFARLAND

Removing Pincushion Distortion in a Cathode-ray Tube

Part 2

Keith decided he would try a more analytical examination of the pincushion problem. "I went to my bookshelf and thumbed through some of my old textbooks. In 'Applied Electronics,' * I found some of the information on magnetic deflection, but it didn't have anything on the pincushion effect, so I went down the hall to the company library. I picked out a couple of books on CRT's and started reading. You almost always have to brush up on your subject before you can get started. You don't just sit down and start cranking out equations. I worked late that Friday night and took the books home with me over the weekend.

"From my review of the CRT fundamentals, I learned that the pincushion effect is caused by the nonlinear relationship between the angular deflection of the beam, which is proportional to the current in the deflection coils, and the Cartesian displacement on the flat face of the tube where the beam hits. The CELCO pincushion corrector was just an extra deflection coil that added fudge factors to the angular deflection of the beam. These fudge factors straightened out the sides of the picture, but they didn't improve the nonlinear relationship between the cur-

* Truman S. Gray, "Applied Electronics" (2d ed.; New York: John Wiley and Sons, 1954). (Text to Keith's introductory course in electrical engineering.)

rent in the deflection coils and the Cartesian displacement. In fact, they made it worse. As I said, it got rid of the pincushion, but we couldn't get the picture to focus. It seemed as if I might do better if I went to the heart of the problem and tried to modify the currents in the deflection coils. This approach had worked on the land-mass simulator, and I thought it might work here too. If I could force a linear relationship between the present coil currents and the Cartesian deflection by putting something else in the coil circuits, I could get rid of the pincushion and still have the resolution we needed."

Keith drew some sketches to make his point clear. "We have a signal coming from the receiving station we want to' display on our CRT. Now let us assume that we used a spherical-face tube [Fig. 1].

"The spot appears where we want it to. The coordinates of a spot on the tube face are linearly proportional to the currents I_x, I_y, which go into the deflection coils, and we get an undistorted picture. Now, if we use a flat-face tube we get the results shown here [Fig. 2].

"The coordinates of the spot are no longer linearly proportional to I_x, I_y, and the spot appears at a distance $\Delta Z = (\Delta x)^2 + (\Delta y)^2$ from where we want it to. Here we have the distorted picture or the pincushion effect.

"I think the best way to attack the problem would

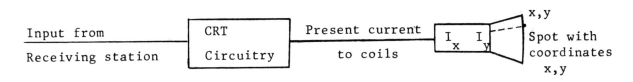

be to operate on I_x, I_y, which might look like this [Fig. 3].

"Here I'_x, I'_y are not linearly proportional to x, y, but I_x and I_y, which represent the true picture, are linearly proportional to x, y.

"I still didn't know exactly where I was headed at this point [early Saturday morning]. I felt I wanted to correct I_x, I_y, but didn't know just what correction factor I would need. It seemed that if I could express the correction factor mathematically, I would be able to mechanize it. I had worked with mechanization problems before on other projects, and I was familiar with some of the techniques. By mechanize I mean design a circuit that takes I_x, I_y in one end, operates on it, and gives I'_x, I'_y out the other."

Keith said that to find the correction factor he needed, he went back to reviewing magnetic-deflection principles for CRT's. "I knew from my reading that the deflection in a spherically faced CRT was proportional to the current in the coils. The relationship was given to me in 'Applied Electronics' as $\sin \phi_y = kI_y$ (where ϕ is the deflection, I is the coil current, and k is a constant). I drew myself a rough sketch to see what this told me."

Keith said he found that a current I_y in the y deflection coil caused the electron beam to be deflected at an angle ϕ_y. The sine of the angle of deflection was directly proportional to the current in the coil, or $\sin \phi_y = kI_y$, where k is a constant, de-

termined by the characteristics of the tube. From his diagram, he saw that the y position (Cartesian coordinate) of the spot was equal to: $y = r (\sin \phi_y)$, or since $\phi_y = kI_y$, $y = rkI_y$. Because of the symmetry of the spherical face, the same relation is true for the x coordinate, or $x = rkI_x$. So the deflection was directly proportional to the current.

Then he sketched the case for the flat-faced tube to express the deflection coordinates x, y, in terms of the current in the coils, just as he had done for the spherically faced tube. He reasoned as follows: He had a current I_x, I_y, which was supposed to cause a deflection of exactly x, y. Instead, it caused a deflection of x, y, $+ \Delta z$. What current then, was necessary to cause a deflection of *exactly* x, y? This relation had been determined for the spherical case as $y = rkI_y$, $x = rkI_x$.

"The case for the flat-face tube was a little harder," Keith continued, "but from the geometry and the deflection principles, I finally arrived at the relation (where θ is the polar coordinate angle or argument in the x, y plane of the tube face): $x = r \sin \theta \tan \phi$. I knew $\tan \theta$ and ϕ in terms of the currents I_x, I_y, in the deflecting coils.

$$\text{Tan } \theta = \frac{I_x}{I_y}, \qquad \sin \phi = k\sqrt{I_x^2 + I_y^2}$$

All I needed now was the relation between $\sin \theta$ and $\tan \theta$ and the relation between $\tan \phi$ and $\sin \phi$. I

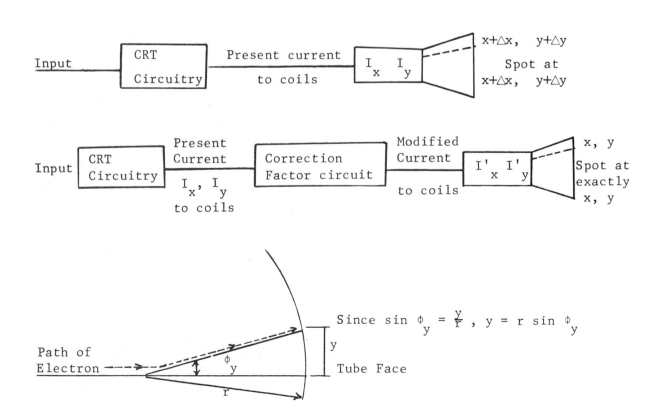

started thumbing through my handbook of math tables and found what I wanted:

$$\tan \phi = \frac{\sin \phi}{\sqrt{1 - \sin^2 \phi}} \qquad \sin \theta = \frac{\tan \theta}{\sqrt{1 + \tan^2 \theta}}$$

Keith indicated that this gave him:

$$x = \frac{r \tan \theta}{\sqrt{1 + \tan^2 \theta}} \qquad \frac{\sin \phi}{\sqrt{1 - \sin^2 \theta}}$$

and substituting $\sin \phi$ and $\tan \theta$:

$$\tan \theta = \frac{I_x}{I_y} \qquad \sin \phi = K \sqrt{1^2 x + 1^2 y}$$

he obtained the relation between the current and the deflection.

$$x = \frac{r\left(\dfrac{I_x}{I_y}\right) k\sqrt{I_x^2 + I_y^2}}{\sqrt{(I_x^2/I_y^2) + 1}\sqrt{1 - k^2(I_x^2 + I_y^2)}}$$

$$= \frac{rkI_x}{\sqrt{1 - k^2(I_x^2 + I_y^2)}}$$

Similarly for the y deflection:

$$y = \frac{rkI_y}{\sqrt{1 - k^2(I_x^2 + I_y^2)}}$$

"This was the expression I was looking for," said Keith, "but it surprised me. According to my equation, the x coordinate was a function of both the current in the x deflection coil and the current in the y deflection coil. I knew what correction factor had

to go into the little black box, but now I wasn't so sure I could mechanize it.

"From my previous experience with mechanization, I knew we could add factors very accurately but we couldn't multiply factors with an accuracy of more than a few percent. This correction factor called for a multiplication of I_x and I_y by

$$\frac{1}{\sqrt{1 - k^2(I_x^2 + I_y^2)}}$$

I didn't think we could generate this corrected current accurately enough to bring the picture within the 0.3 percent distortion specification.

"Assume for a moment that the correction factor is

$$\frac{1}{\sqrt{1 - k^2(I_x^2 + I_y^2)}} = \frac{1}{\sqrt{1 - 0.1}} = \frac{1}{\sqrt{0.9}} = \frac{1}{0.95}$$

This is a pretty fair approximation. The corrected current is thus

$$I'_x = \frac{I_x}{0.95} = 1.05 I_x$$

If we generate this multiplication with an accuracy of 3 percent, then

$$0.03(1.05 I_x) = 0.0315 I_x$$

or 3.15 percent I_x. So we would still have an error of 3.15 percent, and we would still be far from meeting the 0.3 percent error specification."

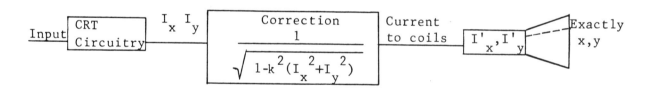

Discussion Questions

1 To what extent do you have the mental programming necessary to generate the same thinking Keith McFarland produced on this project? How close did you come to thinking of his approach? What further programming would you have needed to produce his reasoning and how could you have obtained it?

2 What ideas can you generate for how you might proceed from this point on with Keith's problem?

3 Describe how you would proceed if you were really in his situation.

KEITH McFARLAND

Removing Pincushion Distortion in a Cathode-ray Tube

Part 3

"I had ruled out multiplication as a correction needed to generate an additive correction current." Keith continued, "If we generated the correction current with an accuracy of 3 percent, then 3 percent times 5 percent equals 0.15 percent. Since we could sum the two currents almost without error, the total error would still only be 0.15 percent. Now the problem was to find an additive correction factor that would have the same effect on the current as multiplying by

$$\frac{1}{\sqrt{1 - k^2(I_x^2 + I_y^2)}}$$

Even approximately the same effect would work, because by using addition, there was some extra margin for error. This immediately made me think of infinite series. I knew one could approximate certain functions with the first few terms of an infinite-series expansion. In fact, I could approximate these functions as closely as I wanted by just adding more terms. It is one of the old standby methods that electrical engineers use in circuit analysis. I went back to my math handbook and looked up binomial series. I found an expression that looked something like what was needed.

$$(1 - bI^2) - \frac{1}{2} = \left(1 + \frac{b}{2}I^2 + \frac{3}{8}b^2I^4 + \frac{5}{16}b^3I^6 + \cdots + \right)$$

"If k^2 were substituted for b and $I_x^2 + I_y^2$ for (I^2), the expression fit perfectly. Now I knew I had the problem licked. All that remained was to decide how many terms of the series were needed. The mechanization would be relatively routine. There were still some loose ends to tie up, but I felt that the worst problems were solved. This took quite a bit of the pressure off."

Keith said that the next morning, Sunday, he began thinking about ways to mechanize the terms of

the series he had found. By coincidence it happened that I_x^2, I_y^2, and $(I_x^2 + I_y^2)$ were already being generated for other purposes in the circuits. With the use of log networks, voltage dividers, and summing devices, Keith expected he would be able to generate

$$\sqrt{1 - k^2(I_x^2 + I_y^2)}$$

with sufficient accuracy. The terms of the series, however, required the inverse of this factor, or

$$\frac{1}{\sqrt{1 - k^2(I_x^2 + I_y^2)}}$$

instead of

$$\sqrt{1 - k^2(I_x^2 + I_y^2)}$$

"I began to wonder if the expression

$$\frac{1}{\sqrt{1 - k^2(I_x^2 + I_y^2)}}$$

could be simplified," he said. "The correction factor only had to be generated to about 5 percent accuracy. The term

$$\sqrt{1 - k^2(I_x^2 + I_y^2)}$$

was available almost as a gift from other parts of the circuit. I wrote down the term $1/(1 - X^2)^{1/2}$ and thought about how I might get that denominator into the numerator. I multiplied top and bottom by $(1 + X^2)^{1/2}$. This gave me

$$\frac{(1 + X^2)^{1/2}}{(1 + X^2)^{1/2}(1 - X^2)^{1/2}} = \frac{(1 + X^2)^{1/2}}{(1 - X^2)^{1/4}}$$

"I knew X^2 would be small, so I tried

$$(1 - 0.1)^{1/4} = (0.9)^{1/4} - 0.98 = 1$$

Now I had

$$\frac{(1 + X^2)^{1/2}}{\text{approx. 1}}$$

which was a much simpler expression. This let me use the binomial expansion

$$(1 + bI^2)^{1/2}$$

$$= \left(1 - \frac{bI^2}{2} + \frac{3}{8}b^2I^4 - \frac{3}{16}b^3I^6 + \cdots - \cdots + \cdots - \right)$$

This would be even easier to mechanize. I wrote out the expression making the necessary substitutions.

$$kI_x[1 + k^2(I_x{}^2 + I_y{}^2)]^{1/2}$$

$$= \left[kI_x - kI_x\frac{k^2}{2}(I_x{}^2 + I_y{}^2) + kI_x\frac{3}{8}k^4(I_x{}^2 + I_y{}^2)^2 \right.$$

$$\left. - \cdots + \cdots - \cdots + \right]$$

"I needed to know how many of these terms I was going to have to work with, so I stopped to make some calculations. From one of the sketches I had drawn, I could see the actual deflection coordinates on the face of the CRT were proportional to $\tan \theta$ rather than $\sin \theta$. The error (δ) was simply $r(\tan \theta - \sin \theta)$. I wanted to know how many terms of the series I would have to use. Would the first or at most the first few terms be enough? If so, what kind of accuracy would they give me?"

The display on the cathode-ray tube was designed so that it could be photographed with 70-mm film.

For this reason, only part of the face of the CRT was used for display. [See Exhibits 1 and 2 — Keith McFarland (Part 1)]. Of the area of the face that is used, the centermost section of the face was reserved for actual pictures (the smaller the deflection angle and hence the closer to the center of the tube, the more accurate the picture). To the right of the picture or image frame, there was a data frame with numerical and dot-coded information. The display area then was not symmetric about the center of the tube, and Keith expected separate calculations would have to be made for the maximum error at different edges of the display. The distance from the center of the deflection coils to the tube face along the center line was 8.72 in.

Discussion Questions

1   What is needed next to make Keith's equations describe physical reality? Carry the solution ahead as far as you can.

2   Describe how you would go about proceeding further from the point where you had to stop.

3   What else could you try if Keith's present attempt at solution does not work out?

KEITH McFARLAND

Removing Pincushion Distortion in a Cathode-ray Tube

Part 4

"Before I could determine how many terms of the series were required for 0.3 percent accuracy, I needed a better idea of the kind of error I was dealing with. The image area was in millimeters and my tube data was in inches, so I made some quick conversions with my slide rule." The conversion factor for millimeters to inches is 0.0394. The distance from the center of the deflection yoke to the tube face is 8.72 in. The maximum error on the X axis is on the far edge of the data block.

Maximum X axis deflection = 2.5 in. Thus

$$\tan \phi_x = \frac{2.5 \text{ in.}}{8.72 \text{ in.}} = 0.2865$$

$$\phi_x = 15.99°$$
$$\sin \phi_x = 0.2754$$

max error on X axis = $r(\tan \phi_x - \sin \phi_x)$

$$8.72(0.2865 - 0.2754) = 0.097 \text{ in.}$$

$$\frac{0.097 \text{ in.}}{2.5 \text{ in.}} = 3.9\%$$

Maximum Y axis deflection = 0.9 in. Thus

$$\tan \phi_y = \frac{0.9 \text{ in.}}{8.72 \text{ in.}} = 0.1032$$

$$\phi_y = 5.824°$$
$$\sin \phi_y = 0.10147$$

max error on Y axis = $r(\tan \phi_y - \sin \phi_y)$

$$8.72(0.1032 - 0.10147) = 0.00869 \text{ in.}$$

$$\frac{0.00869}{0.9} = 0.97\%$$

Maximum diagonal deflection = 2.66 in. Thus

$$\tan \phi = \frac{2.66 \text{ in.}}{8.72 \text{ in.}} = 0.3042$$

$$\phi = 17.07°$$
$$\sin \phi = 0.2935$$

max error on diagonal = $r(\tan \phi - \sin \phi)$

$$8.72(.3042 - .2935) = 0.117$$

$$\frac{0.117}{2.66} = 4.4\%$$

"I needed a correction factor of at least 4.1 percent. Since I now had the maximum error for the maximum angle of deflection, and since sin $I = kI$, I merely plugged the sine of the angle into my series expansion. This gave me

$$X_0 = kI_x(1 - \sin^2/2 + 3/8 \sin^4 - 5/16 \sin^6 + \cdots +)$$

or

$$X_0 = kI_x(1 - 0.043 + 0.0027 - \cdots -)$$

"The first term put me well within the 0.3 percent that the specs called for, and I knew I could generate it quite accurately. The second term was only about a twentieth of the first and wouldn't have much effect even if I went to the trouble of producing it. So one term was all I needed to get rid of the pincushion and still meet all the resolution specifications.

"By Sunday afternoon I had a pretty good idea of what the mechanization would be like. I was rather lucky here. I'd done a lot of work with operational amplifiers and log networks before I started on this project. I knew how to use them to get the correction factor I wanted to generate. An engineer with less experience of this sort might have taken a week to do what I was able to do in a weekend. Of course, he might have found another way to do it in 20 minutes; I don't know. I was just glad that I'd had that experience, particularly because time was getting so short. The last part of the problem was a matter of tying up loose ends. All I had to do was to add the proper scaling factors. By Monday morning I had enough of the details worked out so that my technician could begin a breadboard. By Friday we had a working circuit."

Keith added that another nice feature of his solution was that the mechanization employed some standard circuits that Link already had in stock. These circuits were:

1   Adders, which sum input signals into output.
2   Scaling networks, which multiply some input by a constant.
3   Log circuits, which produce an output proportional to the log of the input.
4   Antilog circuits, which are really log circuits used backwards.

The log circuits enable the engineer to get a product of two factors by summing two or more inputs in an adder and then feeding the adder output into an antilog circuit. It is important to note than an input in any of these circuits becomes inverted (the sign is changed) in the output.

Keith's solution is shown in Exhibit 1 — Keith Mc-Farland (Part 4). Pictures of the log and summing networks are shown in Exhibit 2 — Keith McFarland (Part 4). The operational amplifier, symbolized by ▷ is shown in Exhibit 3 — Keith McFarland (Part 4).

Discussion Questions

1   How does a mathematical result like Keith's get translated into physical hardware that people untrained in engineering can solder together to produce the desired result?
2   Compare the difficulties of mathematically modeling a problem like Keith's with that of modeling a mechanical object, such as a noisy faucet, a nut that vibrates loose, or a paper glider that will not fly straight.

Exhibit 1 — Keith McFarland (Part 4)

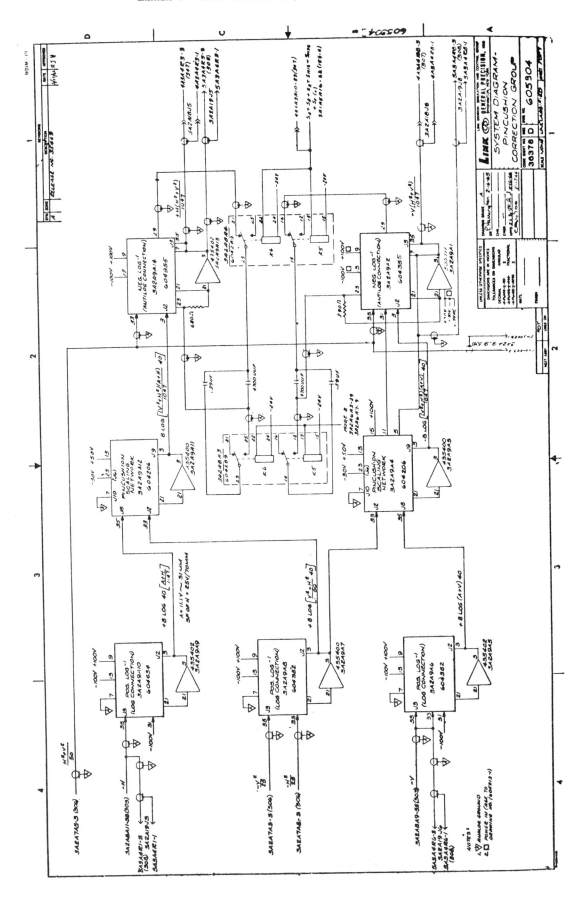